The mutation that made us

Susan J. Lanyon

Second Edition 2016

K-Publishing, Port Huon, Tasmania

Typeset in Computer Modern 11pt using LaTeX.

K
Publishing

Acknowledgements

I wish to thank Anthony Corones (University of New South Wales), who supervised my PhD candidature. Without his support in technical, political and procedural matters, the PhD and this book would not have materialised. Jerry Fodor and Massimo Piattelli-Palmarini kindly examined my thesis and encouraged me to publish my ideas. Thanks to those who challenged and dismissed what I have to say, as they made sure that the arguments are sound and supported by clear evidence. I also would like to thank Peter Krebs for editing and helping out with the translation of some German sources.

Susan Lanyon

Contents

1. Introduction

1.1 Cognition and Evolution

Scholars of Darwin's works will be familiar with the term *natura non facit saltum* – nature does not make leaps. Darwin uses this term several times in *The Origin of Species*. He believed that evolution proceeds gradually due to the fittest of any generation passing on their genes to their offspring, and eventually producing a new species. Much of the current work in the Cognitive Sciences touches on the question of what makes humans unique, and how our complex traits have evolved. There is no doubt that the mainstream view of the evolution of human cognition is one of gradual development (hereafter, 'gradualism'), but I join with the growing number of scholars who find gradualism untenable, and are turning to saltationism as a more plausible approach to account for the evolution of both human anatomy and cognition. I have yet to find an account of human cognitive evolution, however, that takes the saltationist approach to the point of arguing, as I do, that all of the unique cognitive traits attributed to humans arose as the consequence of one crucial mutation, which radically altered the architecture of the ancestral primate brain. Hence I propose that *natura facit saltum*!

Looking at the available evidence, the evolution of both hominid anatomy and human cognition is more satisfactorily accounted for under a saltationist approach. The really important changes in the evolutionary development of hominids, as evidenced by the paleontological record, can be explained by non-adaptational processes, and natural selection does not, and cannot, explain

the sudden emergence of these developments. I further contend
that modern humans[1] did not evolve as a gradual 'trend' toward
becoming 'human' after the split from their Great Ape ancestor.
Most of the anatomical changes that have taken place in our
hominid ancestors (evidenced by the paleontological record) are
merely variations of the regular Great Ape *Bauplan*. Many of
the anatomical traits that we use to define 'human' can already
be found in the fossils of extinct primates. This fact offers a
strong argument against 'special pleading' for proposals that
attempt to link anatomical changes with cognitive advances
in our hominid ancestors. Anatomical traits, like bipedalism,
hands with precision grip and variations in dentition, can all be
the outcome of minor changes in developmental processes that
have emerged suddenly, and have been followed by millions of
years of stasis.

Many scholars have proposed that traits like bipedalism, a
large brain, language, social complexity, tool use, and other
characteristics of humans, must have evolved in tandem during
the evolutionary path of our hominid ancestors. We find view-
points suggesting, for example, that bipedalism is the key to
understanding human brain evolution (Falk, 1992). Considering
that nowhere else in the animal kingdom do we link cognitive
sophistication with the mode of locomotion, why should we
accept this argument here. Witness the bipedal kangaroo as a
case in point!

In addition, it is often purported that the human vocal tract has
evolved gradually, having been fine tuned over perhaps millions
of years, serving an ever increasing complexity in communication
in the form of articulate language (e.g., Ploog, 2002; Pinker,
1994). Using evidence from paleontology, together with the vast
amount of contradiction we find when we look at the acquisition
of sign language, I challenge these presumptions.

The changes in pre-human anatomy seem to have contributed
nothing toward the unique aspects of human cognition and

---

[1]Wherever I use the term "modern human", I am referring to the species
Homo sapiens.

language. However, a major change in the anatomy of the skull of our immediate hominid ancestor, which appears to have emerged suddenly around 120,000 years ago in Africa, provided the fortuitous structure to house the radically altered architecture of the human brain. Other changes at this time, linked to the restructuring of the skull and face, gave rise to the human vocal tract, which was co-opted for the output of our newly emerged language faculty.

An important focus of this book is on the developmental changes, as evidenced by the fossil record, that have led to the crucial changes in our evolutionary past. The most important developmental change was the sudden transformation of *H. erectus* into *H. sapiens*. Several archaic *Homo* species have been proposed as our hominid ancestor (e.g., *H. heidelbergensis*, *H. helmei*, *H. rhodesiensis*), but I prefer to retain the taxa *H. erectus*. I see no reason to grant a special status to any one of the sub-species of *Homo* mentioned, as we have no reason to suppose that any one of these groups had made any advance in cognition from the time of emergence of the genus. Also, recent important fossils of *Homo erectus* found in Dmanisi, Georgia confirm that this species exhibited wide morphological variation within the same population (Lordkipanidze et al., 2013). According to the lead scientist, Lordkipanidze, the fossils representing five individuals, although differing greatly in skull size and shape, nevertheless show no more variation than that found in today's modern populations of humans, as well as chimpanzees. Moreover, these 1.8 million year old fossils bear affinities with many *Homo* fossils found in Africa, and a number of scholars are now debating whether the previously named separated *Homo* species should be moved back into one species, *Homo erectus*.

The saltational evolution of modern human anatomy was due to a neotenous retention of our immediate *Homo erectus* ancestor's early developmental physiology. Modern humans have reached the adult stature of our immediate *H. erectus* ancestor, but have more or less retained the infant proportions of the cranium and face of this ancestor, and I present evidence that supports this major change in our developmental trajectory. A crucial change in the developmental processes of the primate brain, which catapulted *H. sapiens* into a radically different cognitive realm, was part of this neotenous mutation.

I follow many scholars who find evolution by gradual and incremental change, 'driven' by natural selection, problematic for the understanding of the evolution of species (e.g., Bowers, 2006; Chomsky, 2005; Crow, 2002; Eldredge, 1996; Fodor, 2008; Mendíl-Giró, 2006; Gould, 2000; Jenkins, 2000; Lovejoy et al., 1999; Maresca and Schwartz, 2006; Piattelli-Palmarini et al., 2008; Quartz, 2003; Ramus, 2006; Reid, 2007; Schwartz, 1999; Tattersall, 2006). Although I reject the gradualist approach to the evolution of hominids, I accept the tenet proposed by Penn et al. (2008) that "there are no unbridgeable gaps in evolution". The biggest challenge according to Penn is to explain "how the manifest functional discontinuity between extant human and non-human minds could have evolved in a biologically plausible manner". I believe that these unbridgeable gaps can be filled by looking at the developmental changes that have occurred during the span of hominid evolution.

We should also not forget the role that natural hybridization has played in our apparent mosaic of evolutionary change. A single mutation can easily be incorporated into a population by cross-breeding with the parent population to produce novel phenotypes in just one generation (Ackermann et al., 2006). For example, Chouard (2010) reports on the amazing variations of pelvic spines in stickleback populations, as a result of the experimental interbreeding of fish from the sea north of Vancouver and those that live in the bottom of Lake Paxton in British Columbia. The former population is endowed with heavily armoured, sharp pelvic spines, while the latter are basically devoid of spines. The viable hybrid offspring of these 2 populations are endowed with pelvic spines of widely varying sizes. This outcome is due to DNA segments in the *Pitx1* gene that are particularly prone to deletion, and can be shown to be responsible for two-thirds of all variations in pelvic spine length in sticklebacks fished from all over the world's populations. Some observers might consider the large pelvic spines as "monstrous", but nevertheless "hopeful". Richard Goldschmidt coined the term "hopeful monster" in the 1940's to describe evolutionary macromutations that produced novel phenotypes, and these "monsters" may have suddenly emerged as a new species.

It is more than likely that the effects of hybridization produced the various morphologies of the *Homo* genus. According to Duret (2009), the mutation rate in males is higher than in females due to the many more cell divisions in the male germ-cell. In apes, the substitution rate is two times higher on the Y chromosome than the X chromosome.

In the following section, I will outline why the saltationist approach is the only viable option for the understanding of the evolution of hominids and human cognition, including language.

1.2   The Saltationist Approach

The saltationist approach to the evolution of species is concerned mostly with the processes that have led to the emergence of novelty without appeal to gradualism or adaptationism. The emergence of human cognitive complexity was a recent phenomenon in hominid[2] evolution, and the only important trait that separates *Homo sapiens* from the rest of the hominid clade is our radically altered brain. This alternate picture to

---

[2]The term hominid has usually been used to refer to all humans and their ancestors after the split from our Great Ape ancestor (Schwartz, 1999). The term, hominin is now used to refer only to the *Homo* subset, although some scholars include the australopithecines in this group. My preference is to retain the use of the term hominid for all of our ancestors after the split from our Great Ape ancestor. This classification, I believe, should stand, because anatomical and behavioural similarities link all pre-human (non-*Homo sapiens*) hominids.

gradualism is one of discontinuous evolution, where the really important changes in phenotype have been the result of major changes in developmental pathways, which have led to emergent novelties. This book focuses on evidence, primarily from paleontology, archaeology, genetics and the epigenetic processes of development, to argue that modern humans are not the end result of millions of years of gradual and cumulative evolution. Rather, hominids evolved in a saltatory fashion, involving only three major steps. From a developmental perspective, we see that hominid morphology has evolved through the process of heterochrony[3], mostly through the displacement of life-stages, including the retention of infant or juvenile growth patterns (i.e. neoteny). It should be noted that saltational change is an actual observation from the fossil record. Nevertheless, we need to find an explanation that includes all of the factors, both genetic and epigenetic, that led to any saltational event. It is accepted that a saltationist approach bears the same burden that Darwinian incrementalism must face, that is, the problem of reverse engineering in order to extract the underlying causes of change, both in space and time.

1.2.1 Evolutionary Precursors

While recognizing the importance of finding precursors of evolved traits, I give equal credence to theories that point to novelty that comes from the emergent qualities of hierarchical complexification. Evolutionary novelty can arise by other means than the slowly operating 'forces' of natural selection 'building' on pre-existing structures. Schwartz (1999) views natural selection as a process that may prevent an injurious mutation from promulgating through a population, but, as he correctly points out, all this process does is leave room for the fittest, it does not produce change. It is within the developing organism

---

[3]Heterochrony is defined as changes in phenotype through shifts in developmental timing for features already present in ancestors (Gould, 2000).

that we must look to find the important mechanisms of change and the emergence of novel features. Reid (2007) reminds us of the important point, that hierarchical structure, with its emergent levels, is rebuilt in every generation of all organisms. Just as novel emergent qualities arise from the organization of simpler physiological structures, Reid believes that the same process of emergent complexity can be related to the evolutionary past of most organisms, where novelty has appeared at critical threshold points. The upshot here is that many of our newly evolved cognitive traits may not actually entail any identifiable precursors. Later, I outline research pointing to the underlying developmental changes that seem to be responsible for the radically altered architecture of the human brain that supports modern human cognition. It is highly tenable that modern human cognition arose with one crucial mutation that radically change the brain's architecture.

Penn et al. (2008) point out that the dominant tendency in comparative cognitive psychology has been to follow Darwin's belief in continuity between human and nonhuman minds. Accordingly, it is often proclaimed, following Darwin, that the differences are one of degree and not of kind. However, Penn et al. reject Darwin's idea and argue that "the profound biological continuity between human and non-human animals masks an equally profound functional discontinuity between the human and non-human mind". They also think that the discontinuities between human and non-human minds "run much deeper than language or culture alone can explain". Human minds, they hold, have evolved cognitive architectures that can deal with the systematic, hierarchical and relational capabilities of a physical (biological) symbol system[4]. I agree with these authors,

---

[4]There is, of course, the unresolved question whether a physical symbol system is, in fact, sufficient for an 'intelligent' mind. Dreyfus (1979) is skeptical of the view that human lives, which are strongly embedded in changing environments, could ever be described in finite terms, such as symbol manipulating systems require. Nevertheless, he believes that just as planets are not solving differential equations, or following any rules at all, as they swing around the sun, their behaviour is nonetheless lawful.

but take the discontinuity argument even further. I argue that
not only is there a profound discontinuity between human and
non-human minds, but also an equally profound discontinuity
between ancestral hominid and modern human (*Homo sapiens*)
minds.

Bolhuis and Wynne (2009) think that we have tended to assume
that species with shared ancestry are also likely to have similar
cognitive abilities, but things aren't so simple. They believe that
we all too often slip into anthropomorphic language, especially
when crediting human-like traits to other primates. For example,
it has been 'experimentally shown' that capuchin monkeys have
a sense of fairness. They refuse a slice of cucumber when they
see another monkey receiving a preferred grape as a reward
for performing the same task. Just as humans expect to be
rewarded equally for performing similarly, it is *assumed* in this
experiment that monkeys will suffer the same indignation and
reject an inferior reward. However, it has been shown that,
even under normal circumstances, the same monkey will still
refuse the cucumber if it can see a preferred grape that has
been placed in a nearby cage. Similarly, chimpanzees have
been attributed with many of the presumed prerequisites for
language, but many of such claims lack credibility.

In chapter 7, we find that despite the many years of trying to
find precursors for human-like cognition and language, there
is little evidence to show that our unique cognitive traits are
possible without human-like brain architecture. This is not
to say that certain pre-existing computational mechanisms in
ancestral central nervous systems have not been co-opted in the
process of evolution of the human brain. For example, Wallace
(1989) has put forward a plausible account of a link between the
hippocampus serving as a cognitive-mapping structure essential
for any mobile animal, and its possible exaptation in the human
brain for language development. He has shown that an almost

We could, I believe, extend this metaphor to the workings of the human
brain, even if the success of actually finding these 'lawful' workings has so
far eluded us.

identical developmental trajectory takes place in human infants for mapping abilities and their linguistic analogs. In addition, it seems that songbirds may be able to recognize acoustic patterns, which can be likened to a recursive, self-embedding, context-free grammar (Gentner et al., 2006). My intention here is not to discredit the important task of looking for precursors of human-like cognition and language. Rather, the point has to be made that our hominid forebears had made little, if any, advance over and above those cognitive abilities that are currently attributed to our other primate ancestors.

Not so long ago, Piattelli-Palmarini (1989) thought that new work in evolutionary biology and molecular genetics might herald a conceptual shift in ideas about how language and modern human cognition may have arisen. His hope was that this "new picture" of evolutionary mechanisms would be taken up in the transition from Darwinian adaptationism to more plausible explanations that do not include selection pressures as causal "explanations" for novel traits, especially language and human cognition. Similarly, Chomsky's (1988) belief was that molecular biology would discover more informed insights into how the brain/mind has evolved, without the need for adaptationist stories that offer little explanatory value, and simply displace the problem of how initial mutations arise. Their hopes unfortunately have not come to fruition – the stalwarts of gradualism and adaptationism still reign supreme!

1.2.2 Evolutionary Development

Studdert-Kennedy (2000) correctly points out that "every evolutionary step is a change in development" and "evolution does not drive development; development drives evolution". Developmental change is the *only* "driver" of evolution, rather than the most often proposed "driver" of evolution – natural selection. Another important point is made by Raff (1996), who suggests that any explanation of the evolution of form must not only

consider how bodily form is actually generated, but it must also
account for the underlying stability of design. I believe that the
obvious stability in morphology that we see in most organisms,
that has persisted for often millions of years, goes against the
idea that evolution proceeds by gradual, generational change,
leading to the emergence of new species. Raff points out that no
new body plans have originated since the Cambrian Explosion
around 530 million years ago, but that obviously huge evolutionary changes have taken place within these long-conserved body
plans. Developmental changes have opened new opportunities
for novel mutations to exploit most niches available on Earth.
Recent advances in molecular biology reveal how the mechanisms of gene expression relate to the evolution of organisms
and Raff is confident that this knowledge will transform evolutionary studies. The main point to be made is that organisms
do not gradually adapt to environments, but rather settle in
environments that suit their existing phenotypes. Organisms,
in a sense, 'select' their environments. Darwin (1872) pictured
terrestrial animals, gradually evolving into aquatic animals (e.g.,
seals), but the non-adaptational picture is one of 'sudden' or
'saltatory' mutation due to heterochrony, followed by the fortuitous opportunity for these novel phenotypes (e.g., undeveloped
limbs acting as flippers) to exploit new ecosystems.

I present the case for 'saltatory' evolution in chapter 3. Rather
than concentrating on selection forces, we should heed the
advice of Bowers (2006), who suggests that scholars should "stop
wondering about selective pressures" and consider what kind
of mutations might be involved in forming novel phenotypes.
Bowers puts forward strong evidence that the chromosomal
fusion that separated hominids from the pongid ancestor is most
likely responsible for the sudden origin of bipedality, and many
other physiological traits that define hominids. The human
chromosome 2 has resulted from the fusion of the 12th and 13th
chromosomes that are common to chimpanzees, gorillas, and
orangutans. A concentration of regulatory genes involved in the
development of the vertebral column, pelvis, limbs, hands and
feet, as well as the jaw, are located on the human chromosome 2.

The fusion of chromosomes is likely to have provided the crucial mutation that has altered the developmental cascade early in development in humans, and can explain the sudden emergence of these altered traits in early hominids. Accordingly, the sudden emergence of obligate bipedality does not fit with a scenario of a gradual shift from quadruped to biped, under a scheme of selection pressures. Nor does it fit with the paleontological evidence – there are no intermediate fossils. This is just one example of how a seemingly simple genetic change can cause a developmental cascade of changes that may enhance an animal's relationship with its environment, and moreover, allow it to exploit new ecosystems. Falk (1992) wonders why bipedalism became the "preferred" form of locomotion in early hominids. She adds that the reason why hominids "preferred" bipedalism is the biggest unsolved mystery in hominid paleontology. This is a rather strange statement considering all locomotion is constrained by an organism's anatomy, and hence there is no choice to "prefer" an alternative. Moreover, the emergence of bipedalism is certainly not a great mystery, as noted above, and outlined in more detail in chapter 4.

Rosselló and Martín (2006) remind us that we have a much greater understanding of evolutionary mechanisms than in Darwin's and his contemporaries' day. This greater understanding of the underlying processes of evolution, including advancements in paleoanthropology, comparative work in animal communication and complexity theory, they believe, should lead to more sustainable theories for the evolution of species.

Saltationism has unsurprisingly attracted much controversy. Probably one of the most vocal of evolutionary theorists against the value of a saltationist approach to evolution is Richard Dawkins. To illustrate his point, he uses the example from laboratory experiments with flies, which have, in just one generation, their antennae mutated into legs. He claims that these sorts of mutations would, in the main, not be evolutionarily viable (Dawkins, 1991). In his most acerbic attack he likens evolutionary saltation to a "watered-down" form of "creationism".

According to this accusation, those who suggest that viable
mutants (creations) can arise suddenly in the biological record
must be leaning toward 'creationist' doctrine. This accusation arises quite often. For example, Margoliash and Nusbaum
(2009) mischaracterize saltationist views as having resulted from
a failure to find an adaptive explanation for complex traits, like
echolocation and recursion–traits that are often deemed to be
of an 'all-or-nothing' nature. More troubling is that they see
saltationist views as "potentially dangerous in unintentionally
having appeal for creation pseudoscience". Surely not a reason
for anyone to abandon their commitment to a theory.

In chapter 4, I argue that the example by Dawkins of the
sudden mutation of flies within a single generation is exactly
the sorts of mutation that a saltationist would expect for the
production of novel phenotypes. Of course, not every mutation
is an evolutionary success, but occasionally, one arises that
allows an organism to explore and exploit new environments and
resources. In this event, animals do not 'adapt' to environments,
but emerge 'pre-adapted' to 'select' new environments.

Nevertheless, Dawkins (1991) believes that no "sane" modern
biologists are saltationists, and few believe in "sudden" leaps
from one generation to the next . Rather, for Dawkins, "sane"
biologists believe in natural selection as the only possible "creator" of complex organisms. Even the legendary early twentieth
century evolutionary biologists do not escape Dawkins' scathing
remarks. He responds to the "mutationists" (historical rivals of
Darwinism) with "mirth" at the very idea of mutation being a
creative force, leaving natural selection a minor "weeding out"
role in evolution. The notion that the first human brain, that is
twice the size of its father's, or chimp-like brother's, could have
arisen suddenly, Dawkins puts down to "extreme" saltationism,
and is equally dismissed as unbelievable. He characterizes a
saltationist as believing that the eye sprung into form in just
one mutation, or believing that *H. sapiens* emerged from a
"sloping browed" australopithecine in one generational leap. Of
course, this caricature of a saltationist or mutationist is just

as farcical as the caricature of a gradualist believing that the evolution of organisms must occur by gradual change in *each* and *every* generation.

Of course, sometimes we find theories of evolution couched in terms of gradualism that really reflect a case of saltational evolution. For example, (Lee et al., 2014) have traced what they believe is the evolutionary path of modern birds from their dinosaurian ancestors (theropods). Confusingly, they state that birds have evolved "little by little" over 50 million years, but at the same time they show that this evolution was achieved through "12 discernible steps". To my mind, little-by-little implies gradualism and generational change, but if the authors can determine 12 distinct changes over the vast period of 50 million years, this is hardly gradualism. Their analysis uses Bayesian inference to, in effect, smooth the curve of the evolutionary process. This method of analysis seems to me highly inappropriate, considering all mutations are by default random and cannot be accounted for by a probability distribution.

I argue in chapter 4 that the evolution of hominid anatomy proceeded by saltation, but not in just one giant leap from a chimp-like ancestor to *H. sapiens*. However, when it comes to the evolution of the human brain and cognition, the charge of extreme saltationism by Dawkins (1991) holds true. The cognitive capabilities of *H. sapiens* did in fact emerge in one giant saltational leap as evidenced by the sudden and prolific change in technology, but only *together with* the appearance of fully anatomically modern humans in Africa, around 120,000 years ago.

Developmental control genes, that act early in developmental cascades, can cause radical and abrupt downstream changes in morphology (Gilbert et al., 1996; Gould, 2000; Lieberman et al., 2002; Newman and Bhat, 2008; Tardieu, 1998). Drawing on the theory of neoteny[5] we can attribute many of the physiological

---

[5]Neoteny is the retention of early fetal or juvenile developmental growth

changes that have appeared in our ancestral hominid line to variations of 'normal' developmental trajectories (see Antón (2004); Baab (2008); Marchal (2000); Montagu (1989); Schwartz (1999)).

1.2.3   Paedomorphic Hominids

Thirty years ago, Gould (1977) resurrected the idea that the evolution of human morphology should be assessed under a general appearance of 'retardation' of growth during hominid evolution[6].

In support, Vinicius (2005) notes that primates, in general, have very low growth rates from birth up until the sub-adult growth spurt. This derived postnatal growth rate in primates, with the insertion of a juvenile growth phase, is only about 40% of that in other mammals. Other mammals reach sexual maturity at a much earlier age, and this is the stage where anatomical development ceases. Humans represent an extreme case of a slowdown of postnatal growth rates.

Chapter 4 offers evidence from the paleontological record to support the theory for the 'sudden appearance' of all members of our ancestral hominid line due to variations to the 'normal' life stages of our Great Ape ancestor. We find an example of the evolutionary 'developmental' approach with the first hominid split from our pongid ancestor around 5-7 million years ago. The first hominids arose in Africa suddenly and lived for about five million years before their apparent extinction[7].

---

rates or patterns into adulthood.

[6]Gould (1977) traces the notion of human neoteny back to the 1920's. Louis Bolk had explored the fetalization theory for the evolution of humans, and with the arrival of juvenile pongids in zoos and museums, the striking resemblance to humans was recognized. It was noted that during pongid ontogeny, the brain ceases to grow as the jaws dramatically increase in growth, leading to radical differences in form between the juvenile and the adult.

[7]There is much controversy surrounding the allocation of hominids into different species or even different genera, on which I expand in Chapter 4.

A change in developmental trajectory produced a hominid with a remarkably similar physiology to other Great apes, albeit in juvenilized form (Marchal, 2000). A simple mutation appears to have transformed a ape-like ancestor into a bipedal primate. Its basic anatomy did not change in any major way for the entire period of its existence. Variations in size can be put down to sexual dimorphism, as is found within the rest of the Great Ape family. Early hominids were defined by short stature with limb proportions similar to chimpanzees, but with a more or less bipedal stance (Schwartz, 1999). The earliest fossil find to date, *Sahelanthropus tchadensis* (Brunet et al., 2002), was presumed to be bipedal based on the articulation position of the skull to the vertebral column[8], but it nevertheless had the brain size of an extant chimpanzee. Other fossils dating to around six million years have come to light (e.g., *Orrorin tugenensis* and *Ardipithecus ramidus* (Haile-Selassie, 2001)), which are thought to be possible ancestral candidates for the australopithecines.

The australopithecine, nicknamed "Lucy" (*Australopithecus afarensis*) by its discoverer, was dated to 3.6 million years (Johanson and White, 1979). This and other australopithecine fossils indicate that this hominid was more-or-less bipedal, had thick jaws, large molars, reduced canines, but a stature and brain size similar to a chimpanzee (Leakey, 1992).

Around 2.5 million years ago, *Homo erectus* emerged with a 'below-the-neck' anatomy similar in morphology and size to *H. sapiens*, but with a brain only the size of a human one year old. Although the brain size of *H. erectus* was double that of *Australopithecus*, this increase can be accounted for by an almost doubling in body size (Smith and Tompkins, 1995). *H. erectus* made what many believe to be a more sophisticated stone tool, but appears to have not led a 'life-style' involving

---

[8]We should note that this is not a unique trait for hominids. Ankel-Simons (2000) points out that a South American monkey (the *Saimiri*) has the most central position of the foramen magnum of all primates, including humans, but has a quadrupedal locomotion.

built environments, ritual burials, or any other 'social' artefacts
that define 'human' interaction (Binford and Ho, 1985; Gargett,
2000; Tattersall, 2006).

After approximately 2.5 million years of apparent stasis, in
evolutionary terms, of *H. erectus*, we find the sudden appearance of anatomically modern humans in Africa and the nearby
Levant. The morphological changes that identify the *H. sapiens*
cranium show no overlap with our presumed (adult) *H. erectus*
ancestor. Lieberman et al. (2002) note that the *H. sapiens*
cranium appears to have emerged rather abruptly and not by
gradual change. In chapter 4, I present the fossil evidence
supporting the theory that *H. sapiens* emerged suddenly due
to a crucial mutation in our *H. erectus* ancestor. The retention
of the globular skull of an infant *H. erectus* appears to have
allowed room for a change of development for human frontal
and temporal lobes, essential for the hallmarks of human cognition, like creative thinking, artistic expression, planning and
language. Shortly after anatomically modern humans (*H. sapiens*) emerged, we find artefacts that portray a complex society
relating to a major cognitive advance. We also see the quick
demise of all other hominids in Africa, and later in Asia and
Europe.

The genetic evidence for a recent emergence of a small population of modern humans in Africa approximately 120,000 years
ago adds further strength to the saltational theory. Studies
related to human genomics verify this date for the recent emergence of *H. sapiens* from a small founder population in Africa
(Drayna, 2005). Several scholars have put forward scenarios
of how a mutation in a single individual could spread rapidly
over a few generations (e.g., Maresca and Schwartz, 2006; Crow,
2002).

It seems that a major shift in life-history, involving delayed
maturation and a major increase in longevity, may have contributed to the success of *H. sapiens* (Blomquist, 2009; Bogin,
1999; McKinney, 1998). Older generations can invest time and

resources for the survival prospects of younger generations, and this was likely to have been an important factor supporting the spread of this initially small population of humans (Dean, 2006). Up until the emergence of *H. sapiens*, it seems that old age for hominids was only 35-40 years, which is the same as extant chimpanzees (Stringer and Gamble, 1993). Of the entire sample of Neandertal fossils found to date, only 10 percent reached over 35 years of age.

The sudden transformations that appear in hominid physiology support the saltationist approach for the evolution of the hominid clade. The paleontological, archaeological, and genetic evidence clearly supports this claim. This is not to deny that the course of human evolution has been, according to McKinney (1998), a mosaic of developmental alterations. It is unlikely that there was just a single mutation that allowed for developmental shifts in timing for the anatomical features of our ancestral line. However, it appears that only one critical mutation[9] may have caused the initial separation from our Great Ape ancestor, around 7 million years ago. Following this initial mutation it seems that hominids may have diverged in morphology due to hybridization with their Great Ape ancestor. Recent analysis of the chimpanzee genome by Patterson et al. (2006) have found that the human and chimpanzee lineages initially diverged, but later interbred for approximately one million years. Some of the fertile females may have mated back with their ancestral male populations. Schwartz (1999) notes that chromosomes of widely different lengths and shapes can pair upon fertilization to produced quite different phenotypic variation in the offspring[10]. According to Ackermann et al. (2006), hybridization can often lead to novel genotypes/phenotypes as well as the origin of

---

[9]The chromosome fusion theory (Bowers, 2006) and the paedomorphic retention of pelvic proportions (Marchal, 2000) leading to bipedalism, offer strong support for such a mutation.

[10]Robertson (1916) was the first scientist to discover the phenomenon of chromosome fusion. Named after him, Robertson translocations involve participating chromosomes breaking at their centromeres with the long arms fusing to form a single chromosome.

new species. Hybrid skeletons are often morphologically distinct from their parent species, especially in dentition. This
phenomenon is evident in our hominid ancestors. The earliest
hominids as well as the later australopithecines show highly
variable dentition and are usually classified into robust or gracile
purely on the size of their teeth. In fact, Rightmire (1990) finds
it difficult to assign many of the fossils of the Turkana Basin in
Africa to australopithecine, *H. habilis*, or *H. erectus*, due to the
shared morphologies of the crania and jaws. The main point to
be made is that over this vast period of time, the picture is not
one of a gradual and progressive 'trend' in evolution 'leading' to
humans. We readily accept the diversity found in other species,
but seem reluctant to accept a varied morphology within the
hominid clade (Schwartz, 1999).

It appears that another mutation 'event' led to the emergence
of *H. erectus* around 2.6 million years ago. The evolution of *H.
erectus* mainly entailed an increase in body size with a collinear
increase in brain size (Smith and Tompkins, 1995). The last
and final critical mutation caused a radical change in anatomy[11]
and neuro-development, which catapulted *H. erectus* into an
entirely different arena of cognitive abilities. *H. sapiens* was
born out of this mutation.

1.2.4   Archaeology - Saltation and Stasis

In chapter 5, I present the evidence from Archaeology that
matches closely with the paleontological evidence for the sudden
appearance of *H. sapiens*. Not only do we see a stasis of hominid
morphology in the fossil record, but we also find a lack of any of
the cognitive traits that we associate with human-like cognition.
In fact, there is a complete lack of evidence that hominids, for
the first 3-5 million years of their existence, had produced any
technology more complex than that of extant Great Apes.

---

[11]The resemblance of the human skull and face with that of the infant *H.
erectus* is well documented (Arsuaga et al., 1999b; Lieberman et al., 2002;
Montagu, 1989).

Tool technology (or, rather the lack thereof) matches directly the three sudden evolutionary 'events' that produced today's modern humans (*H. sapiens*).

1. The first roughly knapped chopping tools appeared around 2.6 Mya, and remained unsophisticated for over one million years[12] (Semaw, 2000). Although many scholars believe that the 'discovery' of stone tool technology 'heralded' the emergence of human cognition (e.g., Bridgeman, 2005; Gergely and Csibra, 2005), many others are less enthusiastic about this assumption. For example Ambrose (2001) believes that the assemblages of stone tools at this time reflect a least-effort strategy from the available raw materials. They were most likely placed on an anvil and simply smashed with a hammerstone. There is little indication that this technology represented a stylistic "tradition". Similarly, Gargett (1993) dismisses claims for the arrival of "technology", when the term "breaking rocks" would suffice. He observes that many animals are able to transform their environments (e.g., beavers) using natural materials, but we do not automatically associate these skills with a "fabric of meaning" (Gargett, 1993).

2. The second form of stone tool, known as the Acheulian hand-axe, emerged around 1.5 million years ago. This so-named hand-axe is known for its sometimes symmetric shape (Renfrew and Bahn, 1994). It emerged suddenly and was prolific, but remained unchanged for over a million years (Bickerton, 2002). Many have argued that this stone tool is a product of a "mental template", and therefore implies iconography or symbolic behaviour, or even a language capability (e.g., Bridgeman, 2005; Gergely and Csibra, 2005). I follow the opposite viewpoint that sees the emergence of this stone tool as simply the result of a different reduction process during manufacture (see (Ambrose, 2001; Soressi, 2004; Toth and Schick, 1993)). Its emergence does

---

[12]In fact stone tools found alongside 800,000 year old fossils of *Homo antecessor* in Atapuerca, Spain, resemble the first Oldowan stone tool tradition that arose in Africa 2.6 million years ago (Arsuaga et al., 1999a). This means a stasis of form for nearly 2 million years!

not herald a major advance in cognition in *H. erectus*.

3. The most radical change in tool technology, which involves a new complex assemblage (including other materials like bone and antler) followed the appearance of anatomically modern humans, around 120,000 years ago (Leakey, 1992).

The explosion of modern behaviour at this time is, as Tattersall (2006) notes, "underwritten by the acquisition of symbolic cognitive processes". Artifacts become increasingly complex from 100,000 years ago, and we find the first deliberate ritual burials[13] (Walker and Shipman, 1996; Gargett, 1989) at this time. Unequivocal evidence for the deliberate making of fire, and the maintenance of home bases, is also associated only with the emergence of *H. sapiens* (Bellomo, 1994; Binford and Ho, 1985; James, 1989; Ranov et al., 1995; Weiner et al., 1998).

In chapter 6, I survey some of the many plausible scenarios for how the human brain has evolved its radically different architecture. Evidence will be put forward from genetics and developmental biology supporting the view that modern human cognition is likely to have emerged due to a radical 'rewiring' of brain architecture within our immediate hominid ancestor (Crow, 2002; Maresca and Schwartz, 2006; Rakic, 1995; Ramus, 2006). Longer periods of embryonic development have, according to Dean (2006), extended many phases of human brain growth. Just three or four extra days of embryonic development at a critical stage have allowed for three or four more rounds of founder cell mitosis and the large increase of human cortex and associated brain complexity.

Cáceres et al. (2003) make the interesting claim that the bio-chemical changes in human neural cells enable them to function longer than those in other primates. Gene-regulation changes support high levels of cerebral activity over the significantly longer life span in humans. As noted above, the extended

---

[13]See section 5.4 for a more expansive examination of the evidence for hominid burials.

longevity associated with humans was more than likely instrumental in the success of the initially small population of *H. sapiens* as they thrived to eventually become the only remaining hominid species (Dean, 2006).

1.3   The Evolution of Language

In chapter 7, I put forward evidence from linguistics, which supports the view that the evolution of the language faculty is a recent emergence with little (or no) evidence for any precursors for this trait in ancestral hominids or other animals – it is thereby considered unique to *H. sapiens.* In answer to the 'selectionist' or 'adaptationist' account of a language faculty, Piattelli-Palmarini (1989) makes the important point that many of the underlying pre-requisites for language acquisition are so highly interlinked that it is practically impossible to determine what created "selection pressures" for what. Each system of acquiring pragmatic rules, general knowledge about the world, social skills and an understanding of other people's intentions and beliefs presupposes all of the other aptitudes.

Piattelli-Palmarini (1989) makes the valid observation that the currently presumed selective value of human language and cognition is often cited as a cause or *explanation* for how novel traits have arisen. Little is made of the fact that language is largely under-determined by adaptive pressure and is rather "peculiar" in the sense that it is often non-optimal and non-adaptive. He firmly believes that "adaptationism" is often based on a distorted picture of human biological evolution, but he admits that his position is not representative of the majority view in linguistics and cognitive science. My hope is that this book will help to sway opinion away from adaptationism and toward Piattelli-Palmarini's approach.

I also survey many of the theories claiming to have isolated the selection pressures that may have led to full, syntactical language, and also what may have been a precursor for language in the form of proto-language or proto-gesture. I also present theories claiming what may have initiated the first step from animal calls to human-like communication. Most of these theories are found to carry unwarranted assumptions about the cognitive ability of both our early hominid ancestors as well as our recent *H. erectus* forebear.

An interesting target article by Christiansen and Chater (2008) claimed that language, or more specifically Universal Grammar (UG), could not have evolved through biological adaptation *nor* by non-adaptationist genetic processes. They declared that a "biologically determined UG is not evolutionarily viable". This claim is based on the assumption that "there is no credible account of how a genetically specified UG might have evolved". I have no quarrel with this article as far as the rejection of an adaptationist approach based on the gradual evolution of UG due to some sort of selection pressures operating over time. Although the authors reject a genetically based evolution of "language-specific" components of human cognition, they do however intimate "that evolution may have led to adaptations for certain functional features of language". Their reluctance to escape from the adaptationist perspective is a seeming abhorrence to the idea of "such an intricate biological structure emerging de novo through a single macro-mutation". My aim is to show that the emergence of human cognition and language de novo, due to a single mutation, is indeed plausible.

There is no direct evidence for the time and place of the emergence of a language capability. However, when matched with the paleontological and archaeological evidence, what emerges is a picture of the sudden arrival of symbolic behaviour together with a new way of understanding the physical world, and the causal efficacy brought about by this understanding. By inference, we can argue that language emerged along side these other radical changes in cognitive ability, and only with the emergence of *H. sapiens*.

1.4 Summary

The core argument of this work is that all of the unique traits of modern human (*H. sapiens*) cognition and language arose in one major saltational 'event'. I accept that this is a somewhat radical proposal. However, I believe that a careful excursion through the paleontological and archaeological record leads to only one conclusion. Up until the emergence of *H. sapiens*, hominids were little more than bipedal Great Apes. There is no evidence supporting the idea that hominids evolved gradually as a 'trend' leading to humans.

Before the sudden emergence of *H. sapiens*, we find only two hominid speciation events. Many of the physical traits that separate hominids from the other Great Apes can be accounted for by simple changes in developmental timing. Traits like bipedalism, changes in dentition, especially the reduction of canine size, can be shown to be the result of minor deviations in developmental pathways during early development. The first mutation that separated hominids from their primate ancestor appears to have been the result of chromosomal fusion, which led to bipedalism and a change in dentition. The second major change led to the emergence of *H. erectus*, with a much larger body size. *H. erectus* had a post-cranial anatomy that was very similar in size to that of modern humans[14]. The major modification that produced modern human anatomy emerged in just one crucial change in development of our *H. erectus* ancestor. This 'event', as evidenced by the paleontological record, involved the paedomorphic retention of the infant form of globular skull and flat (non-prognathic) face, a remodeling of the vocal tract, and slight modifications of the post-cranial anatomy.

---

[14]Although there appears to be a difference in the spinal column that may cast doubt on many of the abilities that are unquestionably attributed to this species, namely coordinated running and walking, fine motor control over the hands for making stone tools, and breathing control necessary for spoken language (MacLarnon and Hewitt, 2004).

The only artefact that pre-human hominids produced after around 5 million years of 'evolution' is a stone tool, the most symmetrical of which can be shown to be the result of the restraints of mechanics and the raw materials available. Pre-human hominids did not bury their dead, did not engage in any ritual or symbolic behaviour, did not appear to have an intuition for cause and effect, and did not modify their environments in any way.

The human brain did not evolve gradually or incrementally in size or complexity, but instead arose as a result of just one crucial mutation. Brain size in our hominid ancestors was not remarkable in any way, but merely conformed to the expected ratio of brain to body mass of any other primate. We are coming closer to ascertaining the critical mutation that occurred in the ancestral hominid brain and which led to the sudden appearance of *H. sapiens*. Humans emerged with a radically altered brain architecture, which supported an equally radical difference in behavioural abilities, and part of this radical cognitive change allowed for the emergence of language.

2. Problems with Gradualism

2.1 Introduction

Broadly speaking, gradualism is committed to the view that modern human cognition, including language, evolved incrementally through natural selection, after the split from our last Great Ape ancestor. Implicit in this view is the assumption that organisms continuously change, genetically and morphologically, and that these small changes provide the raw material for the 'sorting' processes of natural selection.

For example, Cosmides and Tooby (1994) bemoan the neglect of cognitive scientists for the inclusion of evolutionary theory within their research into the architecture of the mind. They believe that

> evolutionary rigorous theories of adaptive function are the logical foundation on which to build cognitive theories, because the architecture of the human mind acquired its functional organization through the evolutionary process.

They also audaciously claim that "cognitive psychology has an opportunity to turn itself into a theoretically rigorous discipline" by using evolutionary theory to determine the "adaptive function" of the mind. Throughout this book I highlight the fact that many scholars, like Cosmides and Tooby, seem to be committed to the orthodoxy that there is only one evolutionary 'process'

– gradual adaptation to changed conditions of life _caused_ by
natural selection, eventually leading to a transformation of one
species into another. However, I argue that conceptually, 'selection' should be seen as coming _after_ modifications, and only
then can it act _on_ those modifications.

Darwin's basic tenet that evolution proceeds by descent with
modification is taken seriously by most gradualists, but the
emphasis seems to be on minor changes that allow the fittest
of each generation to move continuously forward in a 'march'
leading toward a greater complexity of the species. Gradualist
views tend to range from ultra-Darwinism _à la_ Dawkins (1991),
where taxa are considered to be in a constant struggle to adapt
to change, to a more toned down version known as punctuated
equilibria (Gould and Eldredge, 1977), the latter which still
incorporates gradual change (micro-mutation), but punctuated
by sudden or saltational change (macro-mutation). Regardless
of the 'unit of selection', the picture is one of continuation,
where evolution merely 'fine tunes' what is already there. For
die-hard gradualists, "it is clear that evolutionary change is
gradual from generation to generation, in full agreement with
Darwin" (Pinker and Bloom, 1990)[1].

Dawkins (1991) takes exception to many of the portrayals
of classic Darwinian selectionists, believing them to be merely
caricatures put forward by their rivals. He is most likely right in
arguing that most biologists, when pressed, would not subscribe
to the theory that evolution proceeds due to minor changes in
each and every generation. However, as Pigliucci and Kaplan
(2006) note, the belief that natural selection acts as a driving
force, gradually pushing populations up into higher levels of
fitness, is often unwittingly incorporated into ways of thinking
about how complex organisms have evolved. Dawkins is no
exception. One of the defenders of modern Darwinian orthodoxy,
he admits that he holds a "dispassionate" conviction toward
the Darwinian world-view, believing that it is the only "elegant

---

[1]In the summary of this chapter, I point out that even Darwin retreated
somewhat from this "ultra-Darwinian" position.

solution" that "could, in principle, solve the mystery of our existence" (Dawkins, 1991). Dawkins understands Darwinism as a theory of cumulative processes involving "sorting" over many generations in succession, that may take thousands or even millions of years to complete. He believes that "evolution occurs because, in successive generations, there are slight differences in embryonic development". For Dawkins, to challenge the Darwinian world-view, where "slow, gradual, cumulative natural selection is the ultimate explanation for our existence", is to "deny the very heart of the evolutionary theory"[2].

The problems with gradualism fall into several categories.

- Gradual change does not explain speciation.

- Natural Selection does not explain how the initial trait to be selected arose.

- Gradual change does not match with the now extensive paleontological record.

- Gradual change does not match with the archaeological record.

- The field of Genetics does not support the claim for the gradual accretion of human cognitive modules.

In this chapter, I address the first two items in this list. The other three items at issue are dealt with in turn within the chapters on Paleontology, Archaeology and Genetics.

I make the case that a neo-Darwinian gradualist approach to the evolution of humans is not only untenable, but also lacks explanatory power. Most theories of evolution relate to the "How? When? Why?" questions for how species change over time (Bickerton, 2005b). The gradualist and selectionist approaches are interested in all three questions, but the saltationist detective has more use for the "How?" and "When?" rather than the "Why?" question.

---

[2]Note that Dawkins refers to Darwinian evolution as *the* theory.

2.2   Darwin's Legacy

Darwin can be absolved for his misunderstanding of many biological processes, but it will be shown later in this chapter that many current accounts of how evolution proceeds are still based on preconceptions and false assumptions found in Darwin's texts. I will also show that antagonism to saltationism can be found in Darwin's work, as he sought to defend his theory of evolution against religious orthodoxy[3].

Darwin was committed to the view that nature is in a constant struggle of organism against organism, which he believed led to the gradual improvement of all living things as they 'adapted' to their changing environments. He also predicted that psychology would be securely based on the "necessary acquirement of each mental power and capacity by gradation" (Darwin, 1872).

Eldredge (1996) believes that Darwin was most likely, as most of us, affected by the prevailing sentiments of the day. A common view in Darwin's time was the belief that all change should not only be gradual, but also progressive. According to Gould, Darwin's original commitment to a theory of evolution by gradual change arose from his reading of the influential geologist Charles Lyell. Lyell's uniformitarianism was committed to viewing the geological record as produced by small-scale changes. His work in geology commanded great respect from Darwin and his scientific contemporaries, and Darwin used the analogy of gradual accumulation of geological strata to underpin his theory for the gradual evolution of organisms. Gould (1994) believes that Lyell left no room for uniqueness of evolutionary events of a large scale, and Darwin followed this way of thinking by being quite antagonistic in his views about the role of mass extinction[4] playing a part in evolutionary change.

---

[3]The 'creation' story in the bible, although implausible, can in a sense be considered the ultimate saltation 'event', a point that many of today's antagonists to a saltational theory of evolution seem keen to make.

[4]We have now found much more evidence that points to the likely causes

> [N]atural selection acts only by the preservation and
> accumulation of small inherited modifications, each
> profitable to the preserved being; and as modern
> geology has almost banished such views as the exca-
> vation of a great valley by a single diluvial wave, so
> will natural selection banish the belief of the contin-
> ued creation of new organic beings, or of any great
> and sudden modification in their structure (Darwin,
> 1872).

Darwin, in his quest to counter any ideas of 'special creation',
was locked into a theory of descent with modification, but
only under the assumption that all organisms would change
slowly and in a graduated manner. He left no room for sudden
modifications,

> [a]s natural selection acts solely by accumulating
> slight successive, favourable variations, it can pro-
> duce no great or sudden modifications; it can act
> only by short and slow steps.

To distance himself from naturalists who believed in independent
creation for all species, Darwin maligned them as "a curious
illustration of the blindness of preconceived opinion" (Darwin,
1872). His rejection of the 'unscientific' theories supporting
'sudden' modification of species is made clear.

> There are, however, some who think that species
> have suddenly given birth, through quite unex-
> plained means, to new and totally different forms:
> but, as I have attempted to show, weighty evidence
> can be opposed to the admission of great and abrupt
> modifications. Under a scientific point of view, and

---

of mass extinctions, due to volcanic eruptions, climate change, or massive
extra-terrestrial bodies hitting the earth. Both Lyell and Darwin were
naturally not aware of these relatively recent discoveries. For example,
although we are not certain of the cause, we know that around 290 million
years ago, 96% of marine life was wiped out in a mass extinction (Gould,
1994).

> as leading to further investigation, but little ad-
> vantage is gained by believing that new forms are
> suddenly developed in an inexplicable manner from
> old and widely different forms, over the old belief in
> the creation of species from the dust of the earth.

Also,

> [s]pecies are produced and exterminated by slowly
> acting and still existing causes, and not by miracu-
> lous acts of creation; [...] the improvement of one
> organism entailing the improvement or the extermi-
> nation of others.

Many years after the publication of his theory of evolution, we
find Darwin still perplexed by evidence in the fossil record of
sudden appearances. Known as "Darwin's abominable mystery"
was the vexing issue of the sudden appearance in the fossil
record of the angiosperms (flowering plants) around 200 million
year ago. According to Friedman (2009) this fact motivated
Darwin to speculate that there must have been a lost island
where angiosperms gradually evolved. We now know that the
sudden emergence of flowering plants was due to a sudden
"genome doubling event", which was followed by an abrupt
proliferation.

It seems as if Darwin, in his zeal to reject any notions of
"creation of species from the dust of the earth", is conflating
this 'unscientific' theory with any idea that speciation could
occur through "abrupt modifications". Nevertheless, Darwin
harboured some doubts concerning the lack of evidence for
gradual change, as we see in the next section.

Darwin had observed that descendant organisms, although
recognizable as members of a particular species, are often found
to carry slight modifications differing from the parent organisms.
These slight variations between generations, Darwin believed,
were the fabric of change necessary for evolution to proceed.
Nevertheless, Darwin was challenged and perturbed by the fact

that the geological formations containing fossils, that had been studied so far, did not reveal an infinitude of connecting links. These links should have been evident if evolution proceeded, in the main, by gradual, transformational change. Hence he asks, "[w]hy does not every collection of fossil remains afford plain evidence of the gradation and mutation of the forms of life?". Rather, he admits that whole groups of allied species appear to have "come in suddenly on the successive geological stages".

The problem for Darwin was that this apparent paleontological record of sudden appearance of species lent support for theories put forward by his creationist antagonists. Schwartz (1999) notes, for example, that Georges Cuvier, the great French comparative anatomist, believed that the geological record with all of its abrupt changes in fossils between each sequence or layer, was the result of catastrophic events like floods, as described in the Bible. Cuvier thought that a divine creator may have sent catastrophic 'events' several times, thereby wiping out most of the earth's creatures and subsequently restocking with a new set of species. Cuvier's intensive studies of the Paris Basin limestones correctly revealed a picture of stasis of species often followed by a complete and sudden turnover of species. Species either completely disappeared or else 'tracked' changes in habitat, such as gradual changes in sea level. The paleontological record appeared to support Cuvier's 'explanation', and as Schwartz points out, this pattern argued against evolution by gradual change.

According to Eldredge (1996), Darwin was perturbed by the geological evidence put forward by the 'catastrophists' like Cuvier, which seemed to support the abrupt turnover of species. Concerning his theory of evolution by gradual change, Darwin thought that the lack of smooth transitions in the fossil record was the "most obvious of the objections which may be urged against it". To meet this challenge to his theory of gradual change caused by natural selection, he maintained that the geological record must be far more imperfect than most geologists believed. For Darwin, the hand of man, the wings of bats, the legs of horses, and the vertebrae of the giraffe, were all due to slow and successive modifications, despite the lack of fossils supporting this assumption. He argued that only a small portion of the world had been explored, and in any case, some species may have changed in form, but some of these forms were not preserved in the fossil record. Darwin was quite correct in pointing out that the fossil record is necessarily incomplete. Most organisms never make it into the fossil record. After death, they are either eaten or destroyed by other physical or chemical agents. It is not surprising that Darwin felt that his theory of gradual change was safe from the 'catastrophists.

Now, with much more analysis of the fossil record, the picture is clearer. According to Schwartz (1999), the "fine link" of transformations that Darwin thought should be evident *within* species is non-existent in both the fossil record *and* the extant record. We may find these fine links *between* species, but this is exactly what defines a species: differences, born out of novelty, that separate breeding populations. Not only are Darwin's "fine links" within a species missing in the paleontological record, but his "weighty evidence" against "abrupt" modification is also missing. As we track fossils up through the strata in a cliff face, we find long periods of stasis, then sudden change, which, as Eldredge (1996) reminds us, is an "empirical reality" that was recognised in the earliest days of paleontology. Any evidence for gradual change within the lineage of a species had not been found in the fossil record. As Schwartz (1999) points out, the gaps in the fossil record that Cuvier invoked to explain the "disjointed succession of life" in the paleontological record, were actually real. Species do not grade smoothly one into another, but rather, novelty appears in a saltationary fashion, seemingly out of nowhere. It was only Cuvier's 'creationist' explanation for the sudden appearance of new species that was unfeasible.

Despite the overwhelming evidence in the paleontological record arguing for saltational change, we find gradualists blaming the gap in the fossil record even today. Unfortunately these theories are more often than not conceived to fit an 'assumed' gap in the fossil record. For example, Dawkins argues that we can now only define species in the fossil record because the "awkward intermediates are dead" and the real picture is one of a "smeary continuum" where a species "turns gradually into a new species" (Dawkins, 1991). Evolution, for Dawkins, is acted out by individuals in shifting populations, rather than through the abrupt emergence of new species.

2.3 Neo-Darwinian Evolution

The Modern Synthetic Theory of Evolution (i.e. neo-Darwinism) is a marriage of population-genetics and the more traditional fields of natural history (Gould, 1994). The aim of the Modern Synthesis, which emerged in the early 20th century, was to render all of the subdisciplines of natural history consistent with population genetics and Mendelian principles of microevolutionary change.

Schwartz (1999) believes that the Modern Synthesis, just like Darwin, was also committed to a theory of gradual accumulation of minor mutations, which would eventually transform an organism into a new species. Darwin preferred small scale change over long periods of time, as natural selection honed organisms until they changed from their ancestral form. His experiments using artificial selection when breeding pigeons were used to support the theory that gradual change could lead to speciation. Mathematically inclined geneticists believed that they could track these minor changes within populations, and building in a factor of natural selection, could also answer most of the questions surrounding evolution, particularly how new species are formed. Population genetics, under the banner of the Modern Synthesis, was committed to a research program of adaptationism involving only the gradual accumulation of minor mutations leading to the emergence of new species.

For Maresca and Schwartz (2006), the rationale for the Modern Synthesis was fundamentally flawed. They point out that original genetic experiments on the fruit fly *Drosophila* did not produce new phenotypes, but only manipulated the frequency of variants. This was not evidence of gradual evolutionary change causing transformation of species. Nevertheless, the reasoning behind the approach was that lineages change due to generational and continuous changes in genes, in line with Darwin's theory of gradualism.

According to Gilbert et al. (1996), the role of embryology, which naturally highlighted developmental change within closely related organisms, was an important part of biological study in Darwin's time. Homologies, like mammalian forelimbs and wings, were recognized by Darwin as representing common origins. Anatomical studies revealed many other insights into the evolutionary history of organisms, and embryology had played a major part in establishing many of these homologies. They suggest that the construction of phylogenetic trees, based on embryology, was dismissed as old-fashioned by the proponents of the Modern Synthesis, who felt that embryonic studies needed to be replaced with what they believed was the mathematical elegance of population genetics. Gilbert et al. think that the dismissal of embryology in the Modern Synthesis may have been due to a threat to the projects that were using the gene as the unit of ontogeny and phylogeny. While embryology considered the evolutionary unit to be a morphogenetic field rather than a gene, studies in neo-Darwinian evolution focused on the study of changes in gene frequency. Embryology was more concerned with gene expression rather than with variations in genetically linked populations. Population genetics, which was more statistical than empirical, assumed that all changes in morphology were the result of the accumulation of small gene changes over time. Complex organs were viewed as the product of minor changes in gene expression, and their study did not include the idea of morphogenetic fields.

With the advent of the discovery of DNA sequences, the situation did not change, according to Schwartz and Maresca (2006). The research involving the molecular basis of genetic analysis took on the gradualist model for how evolution proceeds. The emphasis was now on the gradual accumulation of infinitesimally small changes in molecular structure that, accordingly, had a "stultifying effect" on the approach to how phenotypic change occurs. Gould concurs and believes that

> Darwinian theory has been overly burdened by a rigid insistence upon very slow, continuous, adaptive transformations - an unwarranted extrapolation from directional selection upon a single loci in local populations to the origin of new designs (Gould, 1977).

Accordingly, too much emphasis was placed on the "genes for" notion, where it was believed that there was a one to one relationship between a gene and the proteins it built, and the minor physiological differences between organisms (Schwartz and Maresca, 2006).

However, there is much more to the 'building' of an organism than its DNA. Many of the important changes in evolution involve not only the protein building genes, but also the genes that control or act as "micro-managers" for the expression of these genes (Robert, 2006). Major changes in the genome are often due to the deletion or duplication of whole sequences of genetic material, even sometimes the duplication of the whole genome. Robert prefers the orchestra metaphor, where the developmental 'program' is not solely contained in a genetic 'blueprint', but emerges in ontogenetic space and time as a complicated interaction of genes, cellular organization, spatio-temporal conditions, and even external environmental influences. In the following chapters, I outline some of these developmental changes that are apparent both in the evolution of hominids and the emergence of modern humans. For now, though, I will continue to examine the gradualist approach to the evolution of humans.

2.4   Natural Selection and Complexity

Pinker and Bloom (1990) claim that the only successful account
of the origin of complex biological structure is the theory of
natural selection. "Differential reproductive success associated
with heritable variation is the primary organizing force in the
evolution of organisms". Another gene-centred approach takes
the view that "genes are the means by which functional design
features replicate themselves from parent to offspring, they can
be thought of as particles of design", and selection will "drive
an allele systematically upward until it is incorporated into the
species-typical design" (Buss, 2005).

In a similar vein, Cosmides and Tooby (1994) believe that "the
only component of the evolutionary process that can build
complex structures that are functionally organized is natural
selection", which is "a relentlessly hill-climbing process which
tends to replace relatively less efficient designs with ones that
perform better".

However, we cannot assume, as the gradualist would argue,
that the current function of complex traits has arisen incrementally, 'honed' or 'caused' by natural selection. We know
that each organism has a complicated evolutionary history, but
it is far more important to understand how traits have arisen
through their developmental pathways, and only then perhaps,
to speculate on how these traits spread and were maintained
in a population (Pigliucci and Kaplan, 2006). Eldredge (1996)
rightly points out that the conclusions of functional analysis of
evolved traits are merely "a statement of underlying assumptions brought to research in the first place" using the "black
box" of natural selection to explain the origin of function or
behaviour. Pigliucci and Kaplan (2006) are concerned that any
analysis of fitness is bound to *not* take into account some of the
variables that were part of the "fitness equation" at the time of
selection.

Natural Selection as a 'Force'

The key doctrine that underlays this part of classic Darwinian evolutionary theory is that all changes leading to both complexity and speciation are gradual, brought about by natural selection, and 'driven' by external forces. Although neo-Darwinism has excised strict Lamarckianism from its doctrine, we still find selection 'forces' being suggested as the 'cause' of evolutionary change. For example, Bickerton (2005b) states that only some force, like the pressure to communicate, could have *forced* our ancestors to communicate in a more complex fashion than other animals.

Pigliucci and Kaplan (2006) believe that Natural Selection as a 'force' metaphor, pushing a population up "hills of higher fitness landscapes" is far from being a "trivial semantic quibble". This way of thinking about evolution, with all of its assumptions and metaphors, can dangerously "become internalized habits of thought", and its followers often treat alternatives as minority or exceptional views. A prime example comes from Carroll (2006) for whom Natural Selection is "the most powerful force in the universe", a *force* that derives its *power* from the same mathematical principle as compound interest. He believes that the "small differences among individuals, when compounded by natural selection over time, really do add up to the large differences we see among species". Further, Carroll believes it is crucial to appreciate that selection "cannot act on what is not yet needed". This presumably means that natural selection understands what is *needed*, and asserts its *force* accordingly, a case in point for the above charge of a metaphor becoming a dangerously generalized line of thought.

Unfortunately, today's elementary texts on evolutionary theory still focus on the claim that natural selection is the major driver of evolutionary change. Empirical work on evolutionary developmental change and comparative anatomy are rarely mentioned. For example, *Evolution 101* by Moore and Moore

(2006), aimed at students seeking an introduction to evolutionary theory, contains not one theory of evolutionary change that is not subsumed under a scheme of natural selection as the driving force of change. Allopatric speciation is described as an initial separation of a population, usually entailing geographical isolation, followed by different selection pressures 'driving' the separated populations into different habits of life. They believe that each population will eat different food, deal with different predators and perhaps variation of temperatures, and engage in different sexual display. Reference to the necessary biological change that would underpin an animal's ability to undertake a change in diet, to 'deal with' a change in predation, or for that matter, any change in behaviour, is sadly missing. Classic Darwinian theory based on natural selection rules supreme. The first premise of their theory is that populations produce more offspring than 'needed' to replace the parents. This induces competition and struggle for existence, which means that those with heritable traits that enhance fitness and reproduction move their genetic material into the following generations, and *voilà*, "the resulting accumulation of genetic change over many generations is evolution" (Moore and Moore, 2006). Their central idea is that evolution results from small events acting over long periods and can produce large changes. They brush over the process of macroevolution as being where "microevolution shades into macroevolution". The only room they leave for a turnover of species is that brought about by large forces such as extinction. Within this textbook, it becomes apparent that the authors believe that the only really important 'force' in evolution is natural selection.

Fodor (2008) believes that too many evolutionists refer to the process of natural selection as some sort of agent, although we know that only organisms with minds are agents acting out of their intentions. Neither *Natural Selection*, nor *Mother Nature* (used in the metaphorical sense), have causal powers. Fodor is also suspicious of the tactic of placing natural selection's agency in scare quotes like 'prefers', 'designs', and 'wants', as this cover-up does not absolve the flawed adaptationist's approach. Rather, "there is nothing that natural selection cares about; it just happens". He believes that many adaptationist explanations for how phenotypic traits arise are invariably just *post hoc* historical narratives. Fodor asks that evolutionary theory should at least be able to refer to some generalizations, or laws of cause and effect, and also deal with counterfactuals. The challenge to Darwinian adaptationists is to define these 'laws'. Fodor's point is particularly pertinent to the evolution of hominids, which lived in the same environments as many other animals, but are often purported to have 'attracted' their own set of natural selection pressures. The interaction of each organism with its particular environment is extremely complex, and it is an impossible task to reverse engineer the organism in an attempt to unravel the traits that may have been 'selected' in order for an organism to 'adapt' to its current, or even past, environments.

Ramus (2006) believes that "the evolutionary selection question is fascinating but it is so difficult to answer that it is of little practical consequence". His viewpoint stems from the idea that to posit selection forces amounts to deciphering the "intention" behind the evolutionary forces at work when we know that nature has no "intentions". Accordingly, we would need to determine the various stages of the evolutionary pathway of each function together with the selection value at each stage and the genetic factors underpinning this stage, a virtually impossible task. We would also have to show why the individuals that did not have these "well adapted genes" actually transmitted fewer genes (less offspring). Reid (2007) is also sceptical about allusions to fitness in organisms as this is such a broad concept. Moreover he points out that, in some cases, even the "cunning unfit" pass on their genes. Pigliucci and Kaplan (2006) make a similar point. Just because an individual may be lucky enough to have a trait that enhances its relationship with the environment, this says absolutely nothing about its reproductive success.

2.5   Adaptation to Changing Conditions of Life

Classic Darwinian adaptationist theory is underpinned by the assumption that organisms evolve by adapting to the changing conditions of life. Darwin believed that

> [c]hanged conditions of life are of the highest importance in causing variability, both by acting directly on the organisation, and indirectly by affecting the reproductive system Darwin (1872).

The forces that 'drive' evolution are seen to be external to the organism. This idea for how organisms might adapt to new environments is evident when Darwin writes that

> [a] strictly terrestrial animal, by occasionally hunting for food in shallow water, then in streams or lakes, might at last be converted into an animal so thoroughly aquatic as to brave the open ocean (Darwin, 1872)[5].

Current adaptationists do not deviate far from Darwin's way of thinking about evolutionary processes. For example, Bejder and Hall (2002) believe that "hindlimbs likely *began* to regress only after the ancestors of whales entered the aquatic environment" and "limblessness in most snakes is also associated with adoption of a new (burrowing) lifestyle". These authors recognize the importance of developmental change (particularly the arrest of development in the limb bud stage), but they seem unable to escape the gradualist and adaptationist approach that puts a change in behaviour *before* the physiological changes that necessarily precede a 'fit' with a different environment.

---

[5]In Chapter 3, I present evidence that the first cetaceans (the order of mammals including dolphins, whales, etc.), have emerged suddenly in the fossil record, more than likely due to a major mutation that caused the arrest of development in their terrestrial mammalian ancestor.

In chapter 4, I review some of the speculations for how hominids acquired their bipedalism due to the changed conditions of life, or even worse, as a change in their behaviour. Most of these scenarios are infeasible and more often than not, Lamarckian in content.

Dawkins (1991) also expects that "[t]he environment is imposed, and the species evolves to fit it". He adheres to the idea that we need a change in the environment to 'drive' evolution by natural selection, even in the face of many counter examples. An extant but ancient fish species *Latimeria*[6], that has not changed in 250 million years, is for Dawkins an "extreme" and rare example, and he determines that this species did not change because there is "no natural selection pressures in favour of changing". It is unclear why this particular fish endured no "selection pressures", while other fish species, presumably sharing similar environments, were affected. We find this same style of argument applied to hominids, which are claimed by adaptationists to have "evolved" due to selection pressures, while chimpanzees living in identical environments, escaped these "pressures".

In order to support his adaptationist approach to the evolution of species, Dawkins (1991) has built a model of gradual, generational change with a computer program implementing what he calls, "biomorphs". The computer simulation has built-in, step-by-step, mutations of various parts of the biomorphs, which he believes closely mimics evolution. Any "sudden jumps", according to Dawkins, would not be evolutionary feasible. Of course, if one 'creates' the mutation in the first place, then the resultant viability of the creature is a pre-determined evaluation made by the designer of the experimental mutation. In this case, Dawkins is both the designer of all of the conditions imposed on his "biomorphs" and also the arbiter of what he believes to be a viable mutation. These computer models are reminiscent of Darwin's experiments, where he designed the

---

[6]Latimeria belongs to the large family of fish known as coelacanths.

breeding conditions and then 'selected' the beneficial traits that may have been of use to the bearer of these traits. Schwartz and Maresca (2006) remind us, that in Darwin's time it was important, from an economic viewpoint, to consider the value of 'breeding' plants and animals for their usefulness. Darwin selected the plants or animals that were to have the 'reproductive success' just like Dawkins has 'selected' which "biomorphs" are to move into the next generation.

In any case, unlike Dawkins, I argue that 'sudden jumps' are exactly the type of mutations that support the case for the saltationist approach to the evolution of novelty, and the possibility for the emergence of new species. Reid (2007) makes the interesting point that Darwin was actually 'creating' his novel phenotypes within a very short period of time using artificial selection. Accordingly, Darwin's "natural experiments" revealed just how nature could produce "hopeful monstrosities" instantaneously in just one generation.

We should at this point acknowledge that some external conditions can induce mutation. This however, is not the same as claiming that an organism 'adapts' to the changing conditions of life. Maresca and Schwartz (2006) point out that most organisms cope with normal temperature or other environmental changes in their own regions, but sudden temperature changes can induce heat shock, which in turn can have devastating effects on transcription, and eventually DNA-repair mechanisms. They argue that a newly emergent phenotypic trait can result from an instantaneous event. Quoting the early 1900s plant geneticist, Hugo de Vries, it is agreed that "if a feature doesn't kill you, you have it" (Maresca and Schwartz, 2006). The point being that we don't need the 'force' or 'agency' of natural selection to 'cause' the evolution of novelty.

How Gradual is Gradual Evolution?

Some of the contentious issues in both the 'gradualist' and the 'saltationist' approaches to evolutionary theory may lay within their definitions. For example, we learn from Carroll (2006) how some species of fish, known as icefish[7], appear to have evolved adaptations for living in the coldest parts of the freezing oceans of the Antarctic. This was a three-step process.

(1) Over the last 55 million years, the temperature of the Southern Ocean has dropped from around 68F degrees to 30F degrees, and this has affected the habitats of all fish within the Antarctic region. This region is rich in highly oxygenated waters and nutrients. All fish around 55 million years ago were red-blooded, like most other vertebrates. About 25 million years ago, a simple mutation occurred in the DNA code of an Antarctic fish that allowed for the production of a form of antifreeze in the blood. This mutation was caused by a new stretch of code, which was formed when a small nine-letter piece of DNA code, related to a gene that produces a digestive enzyme, broke away and relocated to a new section of this fish's genome. As a consequence, this Antarctic fish was able to cope with the dramatic fall in ocean temperature by keeping its blood viscous. One can easily imagine how this highly fortuitous mutation would quickly (or perhaps gradually) spread throughout the population.

(2) Around 8 million years ago, another mutation occurred producing what are commonly known as icefish. Icefish are so named because of their very pale, or sometimes even transparent, appearance. They have changed from their red-blooded, but cold-adapted, ancestors, due to the loss of the genetic code for making the protein hemoglobin, which binds oxygen in the blood. Icefish survived due to the evolution of a gene that produces the protein myoglobin, also used for binding oxygen in the blood. This protein is encoded by a single gene in most vertebrates.

(3) It seems that within the last 8 million years, an insertion of five additional letters in this gene has occurred, and this mutation has disrupted the production of myoglobin in some

---

[7]A family of species within the suborder Nototheniodae.

icefish. To compensate for this loss of ability to bind oxygen in the blood, these icefish have ended up with larger hearts and blood volumes than those of their red-blooded relatives.

Carroll (2006) is quite sure that the icefish was "not a matter of instant design" but a series of steps. First we have the evolution of anti-freeze in Antarctic fish, then the loss of a gene involved in making hemoglobin, and in some cases, myoglobin. These mutations were produced by simple and traceable changes in the genetic code of these fish.

Gradualists may view this case as support for their position. I view this case as paradigmatic of staged, or step-wise, evolution. A simple mutation 'event' occurs that more or less compels a population to move into a world of new opportunity, in this case, the exploitation of the rich resources of the Antarctic ocean, a place that excludes their red blooded ancestors. A beneficial mutation might spread gradually through a population, but do we consider this gradual evolution? The evolution of a larger heart, and hence the ability to pump larger blood volumes, may be interpreted as the result of gradual evolution. Alternatively, we could argue that we find no novelty here, just the survival of the fittest fish, that is, those with larger hearts *within existing* populations.

A saltationist would view the evolution of the icefish as a series of speciation events, due to the production of a novel phenotype, which more than likely reproductively isolated it from its non-mutated ancestors because of geographical isolation. This is the line of argument, involving saltational change, that I will follow in chapters 3 and 4.

2.6   Adaptation and Speciation

Darwin preferred small scale change over long periods of time, as natural selection honed organisms until they changed from their ancestral form into a new species. He experimented with pigeons, by artificially selecting those with traits that would be the progenitors of the next generation. These experiments underpinned his theory that gradual change would lead to speciation.

Many naturalists in Darwin's time, and earlier, trusted the work of Linnaeus, who had comprehensively placed most living, and some fossil species, within their taxonomic categories. Eldredge (1996) believes that Darwin had to abandon Linnaeus's species in favour of a system where species would be seen as "in progress". Therein lies the irony for Eldredge: if species are forming by gradual change of a parent form into a daughter form, then how can we define a species as a discrete entity? Furthermore, if species are not discrete organisms in space and time, why, he asks, would Darwin title his book *The Origin of Species*? Why have a theory to explain their origin if they don't exist? Nevertheless, Eldredge argues that although Darwin talked about species, he actually

> neatly excised from biology the concept of species
> as discrete entities – feeling it necessary to do so
> in order to establish the validity of the notion of
> evolution (Eldredge, 1996).

He concludes that Darwin, and many evolutionary biologists who have followed, appear to view species "as routinely evolving themselves out of existence". However, he notes, species do not *gradually* evolve themselves out of existence, transforming "chameleon" like, but instead, they persist in space and time often throughout an interval of 10 million years.

Schwartz (1999) highlights the problem of how to *define* a species under a scheme of gradual change. Today, we recognize a species according to its distinctive traits. If a group of organisms is categorized as a species, it must conform to a rigid set of typical characters. Schwartz points out that Darwinian and Mendelian inheritance relate to the continuation of a species, and guarantees its identification. The issue of speciation is important for understanding how certain structures of an organism have changed over time. Under Mendelian inheritance, we may see the selection of small variances within a population, but we cannot claim that these small changes will ever produce a new species. Stotz and Griffiths (2003) believe that a mere reference to gene transfer and random mutation fails to explain how organisms evolve. They are critical of theories that posit a genetic "program" linked to an evolved trait, as the literal genetic code only builds proteins. For Schwartz (1999), speciation should be seen as a series of discrete 'events', or even just one event, where the change is so radical that it reproductively isolates a population due to a breakdown of mate recognition systems.

Speciation of Hominids

Hominid evolution has been the result of regulatory changes during development, just as it has been for the evolution of other animals. Schwartz argues that we therefore should not expect to find a trail of intermediates of gradual change leading from one morphological state to another. However,

> [b]ecause the role of gradualism in hominid evolution has been a mainstay of paleoanthropology, the literature is replete with scenarios of how and why features changed from an apelike to a human like state Schwartz (1999).

Bogin (1999) points out that speciation often occurs due to an altered rate of growth during a particular life-stage. A slight change in ontogeny can produce flow-on effects on subsequent stages. All life stages of humans have been extended compared with apes, and life history delays have produced overdevelopment, especially of the body size and a correlated larger brain. Our actual rate of growth is comparable to other apes and also our hominid ancestors, but modern humans experience an extended period of growth in each life-stage (McKinney, 1998). The changes in life-history stages of hominids have most likely been the catalyst for reproductively isolating certain groups.

Accordingly, when a dramatic change in phenotypic structure occurs due to developmental change, we should not speak of it as arising *for* anything (Tattersall, 2006). This does not devalue the observations that certain structures that appeared in the hominid line may have had a selective advantage. For example, the earliest hominids show modifications of the functional complex for heavy chewing (McHenry, 1994). These hominids appear to have partially retained the juvenile form of the ancestral ape skull and face (reduced prognathism). The shortened muzzle and palate reduced the size of the canines, but also produced post-canine megadontia. These huge cheek teeth would have allowed for the exploitation of special niches where the diet involved heavy chewing. The acquisition of food is undoubtedly the main selection pressure for the survival of all animals, including humans.

Gould (1989) complains that "most of us labor under a false impression about the pattern of human evolution". The evolution of *H. erectus* is a case in point. *H. erectus* fossils enter the paleontological record by 1.8 million years ago and varied little in morphology or behaviour over a period of nearly two million years of 'evolution', in Africa and Asia, and in widely differing ecological settings. Gould makes the point that it is highly *improbable* that all of these populations, in different continents, gradually moved up the 'ladder of life' leading to the 'inevitable pinnacle', *H. sapiens*. Rather, we see a conservation of form and no evidence of any 'advances' in cognition over this vast period of time.

2.7   Natural Selection as Conservator

In contrast to the Darwinian understanding of the role of natural
selection in evolution, (Groves, 1989) argues that "natural
selection is primarily an agency of conservation, not of change".
This claim is supported by the lack of a definitive example of
any genus that contains a neat trajectory of species leading
from simple to more complex forms. The fossil record shows
rapid speciation 'events' followed by long periods of stasis, often
lasting millions of years, supporting Groves's findings of mainly
conservation of phenotypes in the fossil record.

For Reid (2007) the terms 'natural selection' and 'fitness' have
very little to do with evolution. He rightly points out that
natural selection is "irrelevant as a causal explanation of evolutionary change". He proposes that the only role for natural
selection is in the study of population dynamics. An example is
the field work carried out by the Australian biologists Phillips
and Shine (2006) investigating the impact of the introduction
of an invasive species on native fauna[8]. Toads were introduced
into Australia in 1935 in order to attempt to control insects
affecting sugar cane production in Queensland. A side effect
on native animals was the high toxicity of these toads, which
were to become the prey of Australian black snakes (*Pseudechis
porphyriacus*). Before the introduction of cane toads, snakes
mainly predated non-toxic frogs. However, snakes with large
heads, and thus a large gape, were able to eat a toad, but
just one ingestion of a large toad is fatal. Phillips and Shine
(2006) have found that selection against large headed snakes
has induced morphological change in these snakes. In areas
where cane toads are abundant, there is a tendency toward
small heads and large bodies. Small heads mean a smaller gape,

---

[8]I use this example as an illustration of an approach by gradualists
toward evolutionary theory, but in no way do I mean to discredit this
valuable study on human-induced environmental change to ecosystems. As
Pigliucci and Kaplan (2006) point out, quantitative genetics may be very
useful in making short term predictions for species survival or fitness within
an ecosystem.

which in turn restricts these snakes to prey upon only small toads, and larger body mass allows for a greater capability to cope with toxins. This change in population dynamics is understandable and entirely feasible, but is it evolution? They refer to these changes as "evolved responses", but I argue that this scenario only illustrates a case of population dynamics. Large headed black snakes are still found in all regions where cane toads are found, so we do not have a case for the evolution of a new species. One can easily imagine that the population of black snakes in Queensland could change equally rapidly in morphology, if the cane toad was exterminated. Once again, this would merely be an example of intra-species population dynamics, not 'evolution' *per se*.

Darwinian gradualist evolutionary theory relies on generational change and the accumulation of small genetic modifications within populations. Generational change is based on Mendelian principles of inheritance. Eldredge (1996) reminds us that Mendelian inheritance is more or less particulate, with phenotypes representing alternate states of genes (alleles). Schwartz (1999) points out that Mendelian principles of inheritance may show how natural selection can move a population toward a certain trait, but it cannot produce anything new. For example, if height was being selected for, then the average individual of the following generation may be taller, but this cannot go on indefinitely. The following generations will, on average, not exceed the maximum height of the preceding generations. As noted above with the icefish, those with the larger hearts were able to better cope with the changing environment, but this enlargement could not go on unchecked without a major reorganization of this vertebrate's anatomy. Schwartz (2001) accepts that a faster than average predator may obtain more food, which may lead to increased reproductive success. However, he rejects the assumption that, over time, this predator would change, that is, evolve. All that concepts like 'faster', 'stronger' or 'better' demonstrate, is the survival of a species, but they do not explain an organism's transformation into something else, that is, the *evolution of species*.

2.8   Brain Evolution

The mainstream approach to the evolution of the hominid brain generally postulates some kind of external 'force' driving a gradual increase in brain size and complexity (e.g., Buss, 2005; Calvin, 2006; Cosmides and Tooby, 1994; Schoenemann, 2006; Wallace, 1989). The brain is thought to have evolved gradually over millions of generations since the split of the hominid clade from our Great Ape ancestor. For example, following the continuous, gradualist paradigm for the evolution of the hominid brain, Schoenemann (2006) puts forward a calculation for an increase of around 8 mm^3 in brain size per generation over a 2.5 million year time span to account for the difference in brain size of *H. erectus* and *H. sapiens*. Similarly, Milton (2006) supports Edward O. Wilson's estimation that for more than two million years "the human brain grew by about a tablespoon every 100,000 years". She adds that

> apparently each tablespoonful of brain matter added in the genus *Homo* brought rewards that favored intensification of the trend toward social and technological advancement (Milton, 2006).

Both of these scenarios rely on the theory that the brain has gradually increased by very small amounts generation-by-generation, a highly unlikely scenario.

Committed to the statistical approach, Parisi (2003) believes that his computer simulations of artificial life can tell us a great deal about the evolution of our cognitive architecture. His models however, are based on the assumption that evolution is gradual and that his information-processing modules are direct simulations of biological substrates. Krebs (2007, 2008) illustrates some of the issues of assumptions built into models and the inherent dangers of extrapolating results from computer simulations. The problem for Krebs lies in relating computational or mathematical entities with real biology and the observable 'behaviour' of neurons.

Some adaptationists even attribute some sort of ability for an animal to exert control over the path of their own brain evolution. A case in point is the claim that "organisms that *need* larger brains will only be able to evolve them and hence occupy the new niches they *seek* to occupy" (Dunbar, 2001, my italics). Moreover, "only those species for whom the more developed behavioural process was essential *went to the trouble of evolving large brains*" (Dunbar, 2001, my italics).

Smith and Tompkins (1995) are particularly critical of speculations on this issue on the grounds that these accounts of brain evolution are usually "fossil free". Eldredge is very cautious about theories that assume a gradual linear increase in hominid brain size. He also believes the gradual approach toward the evolution of the brain to be an "a priori extrapolationist view – unsullied by any reference to examples in nature" (Eldredge, 1996). The fossil record reveals that early hominid brains were only slightly larger than those of their Great Ape ancestors, and this slight enlargement can easily be accounted for by a slightly larger body size. Hominid brains 4 million years ago were only 450 cc, and their relative brain size did not increase over the approximately 5 million years of their 'evolution' (Quartz, 2003).

H. erectus arose around 1.8 million years ago in Africa with a brain size between 750 cc and 1000 cc and it remained the same for 1.3 million years, that is, for most of its species' existence (Quartz, 2003; Walker and Shipman, 1996). Once again, this doubling of brain size can be accounted for by a dramatic increase in body size of *H. erectus* over its hominid (australopithecine) ancestor. *H. erectus* had a similar growth trajectory to modern humans 'below the neck', but certainly not 'above'. The average adult *H. erectus* brain size was still only the size of a human one year old and most importantly, seems to have reached its maximum size very early on in its ontogeny, unlike the human pattern of brain growth. Eldredge (1996) quite rightly dismisses theories that argue for an increasing 'trend' in hominid brain size, as the picture is rather one of 'sudden' enlargement, associated with an increase in body mass[9], followed by millions of years of stasis.

Despite an obvious increase of brain size in our hominid ancestors, Raff (1996) warns against viewing the human brain as just a "scaled up" version of the chimpanzee brain. He claims that the main differences are due to profound changes in architecture of the human brain, and that they cannot be attributed to a simple increase in size. Moreover, the correlation of brain size and its surface appearance, and intelligence, can be highly misleading. Klinowska (1994) points out that the brains of several distinguished people, who have bequeathed their brain to science, have been found to be quite ordinary.

Proponents of evolutionary psychology usually view the hominid brain as having evolved over millions of years, building functional circuits, or modules, that relate directly to the adaptive problems faced by our hunter-gatherer ancestors. It is often claimed that changes in hominid behaviour, such as maintaining home bases, manufacturing or using stone tools, hunting, foraging and food sharing are evidence of an advanced brain, but Blumenberg (1983) warns against this proposal as all of these behaviours have been witnessed in extant pongid species. Moreover, as Jerrison (1983) points out, other small brained animals like birds and wolves, as well as social insects, are capable of effective control of complex social and other behaviour. We do not view *their* brains as caught in a feedback mechanism of brain enlargement and behavioural complexity. He adds that, given what we know about a connection with brain size and cognitive ability, we should not view evolution as a following any trends. We should not expect to find a relationship between the brain size of any animal and its intelligence.

Quartz (2003) points out that neurogenesis in the mammalian brain is highly conserved. He argues that if natural selection

---

[9]According to Smith and Tompkins (1995), examination of a *H. erectus* male fossil indicates that it did not experience the adolescent growth spurt that most modern humans go through at puberty. However, it had reached the height of a modern human adult male by 7 years old.

operated on neurogenesis, then we should see more variation in some individual brain structures within different species. Across 131 mammalian species, the size of brain structures, with the exception of the olfactory lobe, are highly correlated. Where we see a difference in brain sizes, it appears to be the result of a global change in proliferative processes, which preserves the linked regularities among the various brain structures. This global change in the proliferative processes results from hete- rochronic changes in the duration of neurogenesis. The spatial organization of all mammalian brains reflects the highly con- served order of neurogenesis under the control of an interacting group of regulatory genes. For Quartz, this suggests that the re- stricted phases of neurogenesis could not allow for independent selective pressures to operate on different brain structures. It is therefore infeasible to divide the neocortex into a collection of autonomous modules that determine human traits that have evolved independently.

Mundale (2003) is particularly critical of models in evolution- ary psychology and their lack of integration into the biological sciences. She takes issue with Cosmides and Tooby (1994), who, she claims, are only interested in what adaptive information- problems our ancestors "evolved to solve". Evolutionary psy- chology in general circumvents implementational issues, which Mundale believes severely weakens their theories. The extremely coarse-grained and highly abstract functional modules proposed by evolutionary psychology cannot be realistically correlated with a fine-grained, physical implementation required by infor- mation processing (IP) systems. Furthermore, Mundale accuses evolutionary psychology of eroding the understanding of IP mechanisms by ignoring the important advances in neuroscience with regard to the role of neurotransmitters and transmitter- specific systems in the brain. By confining the study of cognitive function to psychological modules that fit into an "adaptive" landscape, Mundale rightly concludes that evolutionary psychol- ogy may actually *hinder* progress in elucidating the cognitive architecture of the brain.

Selection Pressures and Brain Evolution

In the following sections, I criticise some of the adaptationist
theories that are claimed to provide the answers for how the
hominid brain has evolved since the split from our Great Ape
ancestor. We will find that all of these adaptationist 'theories' offer little in the way of feasible scenarios for how the
human brain has evolved. Under the banner of Evolutionary
Psychology, many theories are formulated to support a claim
that the brain has gradually evolved modules that have direct
links with the selection pressures thought to have been active
throughout the history of our hominid ancestors. Buss (2005)
is motivated by what he believes is the need for "reformulating
and expanding the social sciences (and medical sciences) in light
of the progressive mapping of our species' evolved architecture".
Buss views the mind as computational, with functional circuits
that have evolved to produce the circuit logic of human nature.
He optimistically believes that Evolutionary Psychology models
should eventually be able to provide a detailed specification
of the underlying genetic and developmental bases of these
mechanisms, as well as the neural architecture that supports
them. For Buss, most functional organization, or as he puts it,
adaptations, must have arisen as the result of selection, which
he believes is the only known natural physical process that
can build this highly ordered architecture of the brain. Naturally, following a "sudden emergence" model for the emergence
of the human brain and human cognition, I dispute Buss's
adaptational approach.

Hunting and other Social Behaviours

Some claim that there is a correlation between behaviour and
the evolution of brain size, which has been brought about by
the ever pervasive 'forces' of natural selection. For example,
Walker and Shipman believe that

> braininess probably coevolved with a predatory be-
> havior and that a relatively encephalized species,
> like *H. erectus*, was almost certainly an effective
> hunter (Walker and Shipman, 1996).

Similarly, Milton (2006) thinks that brain size in the human
line grew as a result of the pressures to acquire a steady and
dependable supply of very high quality foods under environmental conditions in which new dietary challenges made former
foraging behaviors "somehow inadequate". Milton admits that
specialized carnivores and herbivores abounded in the same
environment as hominids, and that they were evolving at the
same time. However, these conditions were only

> forcing [humans] to become a new type of omni-
> vore, one ultimately dependent on social and tech-
> nological innovation and thus, to a great extent, on
> brainpower (Milton, 2006).

Deaner and van Schaik (2001) propose that an increase in
brain size was 'selected' in order to cope with demands of a
species that began to live together gregariously. Individuals
apparently *decide* to start living together, perhaps to avoid
predation. Living together in groups, they claim, would have
meant an increase in social processing demands and perhaps
the need for a greater foraging range to provide enough food
to support the group. This in turn, they think, would have
driven the evolution of brain structures that would have met an
increase in spacial memory demands. It is not made apparent
why this evolutionary scenario only applied to hominids, and
not to, for example, Bonobos, which also live together in similar
gregarious social groups.

A similar line is taken by Tomasello et al. (2005) who have
suggested that the motivation to share experiences and emotions, to plan and collaborate, may have arisen in hominids
when selection pressures forced individuals, or perhaps a group
of primates, to turn their competitive style of behavior into

collaborative and mutually beneficent social behavior. They suggest that competition for limited resources may have driven some primates to modify their behavior and in so doing, has given them an edge for survival. But Bickerton (2005c) justifiably asks, "What scarce resources? What change in ecology? Why did humans, rather than any of the other primate species, begin to collaborate?". He points out that there is absolutely no evidence that any other primate engages in collaborative behavior. If collaboration is possible for primates, and it is such an evolutionary advantage, why do we not find this behaviour among many other primate species?, asks Bickerton.

Cosmides and Tooby (1994) argue that just as the human body has a "different machine" for solving different problems, the mind has evolved specialized modules to solve its processing problems. The heart evolved to pump blood, and the liver is specialized for eliminating poisons. Inspired by the discovery that humans appear to have an innate universal grammar for language, they believe that humans have also evolved a universal grammar of social reasoning. According to this approach, our hominid ancestors developed specialized circuits to deal with social exchange, threat, coalition action, mate choice, and a cheater detection device. For Cosmides and Tooby, paleoanthropological evidence suggests that hominids have been engaging in their 'universal grammar of social reasoning' for millions of years - "more than enough time for selection to shape specialized mechanisms" like, for example, a 'cheater detector' module.

The archaeological record challenges many of the claims made within this section (see chapter 5 for more detail). Firstly, *H. erectus* seems to have been little more than an opportunistic scavenger and not the hunter that so many claim this species to have been (Binford and Ho, 1985). In fact, we find little or no evidence for any advance in hominid 'intelligence' over the cognitive ability of our Great Ape ancestor, until the sudden emergence of *H. sapiens*. Secondly, the assumed increase in encephalization of our hominid ancestors can be accounted for by the increase in body size. We will see in section 6.3 how, generally, mammalian brain size is closely linked to body size. Claims that *H. erectus* was a more specialized hunter than any other predator are unwarranted. Our hominid ancestors did not make fire, did not bury their dead, did not engage in any ritual behaviour, nor were they technologically adept.

Sexual Selection

A large part of social organization in most animals is without doubt based on the interaction of the sexes. Many traits associated with reproduction that we find in modern humans, be they anatomical or behavioural, are claimed to have arisen by the evolutionary process of sexual selection. For example, Buss (2005) is convinced that we have a "genetic blueprint", which has evolved over time and determines modern humans' sexual selection preferences, mate acquisition and parenting behaviours. He believes that by studying the impact of a foraging life-style on the mother-infant bond, he can shed light on how our psychological 'modules' have evolved. However this approach is fraught with problems. We have no idea about the culture or life-style of ancestral females, in fact, we don't even know whether males hunted and females foraged. Pruetz and Bertolani (2007) have studied a group of chimpanzees in Senegal, in which most of the hunting is actually performed by females and juveniles. We cannot make broad assumptions about the gender roles of ancestral hominids.

Despite his caveat that "there are also many features of the ancestral world about which we are completely ignorant", Buss (2005) nevertheless claims that we *know* that hominids had a sexual division of labor involving differential rates of hunting and gathering, had long periods of biparental investment in offspring, and engaged in enduring male-female mateships. However, evidence from the fossil record points to the fact that many of our hominid ancestors had a high degree of sexual dimorphism, which makes it difficult to make reliable predictions about their sexual-social systems. As a result of his studies of sexual and canine dimorphism in fossil hominids, Plavcan (1997) concludes that virtually any sort of mating behaviour could have been in operation, especially with male-to-male competitive situations. Further, it is apparent that hominids had life histories much more similar to apes than modern humans (Tardieu, 1998). They reached sexual maturity at a much earlier time, and females were able to produce offspring throughout their life-span, which was similar to that of extant Great Apes. In addition, as Panksepp (1998) notes, although most mammals share an underlying common limbic psycho-sexual arousal system, they vary widely in terms of their systems of sexual behaviour. We only have to look at our presumed closest 'cousins', *Pan paniscus* (bonobos) and *Pan troglodytes* (common chimpanzee), to see how widely social and sexual behaviours can vary within two very closely related species (Savage-Rumbaugh and Wilkerson, 1978). The external genitalia of *Pan paniscus* is anteriorly rotated, and copulation, including homosexual copulation, takes place throughout their cycle. Feeding and food sharing, together with many other forms of social interaction, is common, and often precedes sexual activity in *Pan paniscus*, but not in *Pan troglodytes*. Bonobo social structure is based on a non-dominance approach, which is quite different from that of the common chimpanzee. Recent studies by Gilby et al. (2010) show that male chimpanzees (*Pan troglodytes*) do not trade meat for sex, and very little interaction exists between the sexes that could be linked to what we think of as social or sexual interaction, other than the copulatory act itself.

The challenge to theories based on an unwarranted presumption of social organization of ancestral hominids is supported by work in the field. Bermúdez de Castro and Nicolás (1997) have excavated a large sample of hominid remains representing 32 individuals belonging to the same biological population from the Middle Pleistocene (around 780,000 years ago) in what is now Spain. These hominids are presumed to be an ancestral population to European Neandertals. Their life-span was no

longer than 40 years, and the mortality pattern was quite different from modern day foraging groups. The low survivorship of
infants and children (drastically under-represented in the fossil
remains) together with a high mortality rate of adolescents
and young adults presents a puzzle to these palaeontologists,
who argue that this would not have been a viable population.
They suggest that some kind of different behaviour may be
responsible for deviations from 'normal' life-history profiles and
mortality rates, especially that found in young adults where
53% of individuals, mostly females, died between 15 and 18
years of age and six individuals died at a similar age, between
14 and 15 years. The high mortality rate of young females may
explain the unusually small amount of infant remains due to the
non-parental care of orphans. Further differences in behaviour
can be derived from the fossil evidence. Although the sex ratio
for this population was 1:1, there appears to be a high degree of
sexual dimorphism. Analysis of the mandibles (jaw bones) and
dentition show a clear pattern of large size difference between
males and females. This is a clear deviation from the regular
pattern found in most primates where large sexual dimorphism
usually means a large departure from a 1:1 distribution of the
sexes in a given population.

A further problem noted by Eldredge (2003) is the assumption
that all forms of sexual behaviour in humans is about reproduction, and that all behaviour reflects a competitive race to pass
our genes to the next generation. He notes that Bonobos engage
in sexual liaisons that often have little to do with reproduction.
Caporael is also wary of generalizations that equate complex
traits with genetic fitness.

> [T]o support the claim that a male preference for
> young and attractive females is a genetically based
> result of evolution by natural selection, we would
> have to show that such picky males outreproduced
> males willing to take any possible copulatory op-
> portunity in the small hunter-gather groups of the
> evolutionary past (Caporael, 2003).

To posit genes for various dispositions may, according to Caporael, be interesting for popular culture, but this can also lead
to "harmful fictions", and when incorporated into evolutionary
theory, can easily become a discipline of "myth" rather than of
science.

Climate Change

A basic tenet of Darwinian gradualism is that organisms become
somehow 'adapted' to their environments. As ecosystems or
environments change, organisms are assumed to change in order
to 'fit in' with those changes. Individuals within a population
that do not measure up to the new pressures are assumed to
be 'selected' against, or 'weeded' out of the gene-pool. Climate
change is often put forward as a mechanism that drives the
'need' for change in organisms.

During the evolutionary history of our hominid ancestors, there
were many climatic fluctuations. Calvin (2006) points out that
Africa was cooling and drying 4 to 6 million years ago, but
around 2.5 million years ago, the earth experienced abrupt
and frequent cooling episodes. Entire forests would have disappeared over very short periods of time. He rightly adds that
the evolution of anatomic adaptations in hominids could not
have kept pace with these abrupt climate changes, some which
would have occurred within the lifetime of single individuals.
Despite his conclusion that hominids could not have adapted
to these dramatic fluctuations in climate, he strangely makes
an exception for their brains. Following a gradualist approach
for the evolution of the brain, he suggests that "still, these environmental fluctuations could have promoted the incremental
accumulation of mental abilities that conferred greater behavioral flexibility" (Calvin, 2006). The fact that other animals
in these changing environments did not 'accumulate' greater
mental abilities renders Calvin's claims infeasible.

Eldredge (1996) considers theories of evolution relying on climate change as the 'selecting' agent to often be a grossly over-

simplified and distorted view of how organisms interact with their complex ecosystems. Organisms are immersed in their ecosystems and only manage to flourish if they can sustain a complex flow of energy within their stable surroundings. He finds that when ecosystem changes, organisms tend to either engage in "habitat tracking", or otherwise suffer extinction. They cannot miraculously change their phenotypes, which of course includes their brains and behaviour, to match a changed habitat. He is also troubled by traditional accounts of how evolution "springs into action" whenever or wherever an empty niche becomes available. The adaptationist approach seems to assume that organisms will miraculously mould themselves to suit changed environmental conditions. Eldredge's "habitat tracking" model, where species migrate to and from ecosystems within which they can sustain a viable existence, is by far the more feasible scenario.

As more hominid fossils are discovered, it becomes increasingly apparent that the first hominids emerged in environments that were similar to where we find extant chimpanzees (Pi et al., 1997). It therefore seems improbable that environmental pressures were responsible for the evolution of the initial stage of the hominid clade. Moreover, studies of endocasts reveal that the first hominids, that emerged around 5 million years ago, had ape-like cerebral cortices and retained this pattern right up until later finds of 1 to 2 million year old fossils. The often uncritically accepted view, that a climatic change of increasing aridity forced our primate ancestor out of a forest environment onto the savanna and into bush ecozones, has been challenged by Blumenberg (1983) and Pi et al. (1997). The Pliocene sites, where the remains of small-brained hominids have been found, indicate dry environments with varying proportions of forest, woodland, bush, and savanna. The climate in these regions of Africa changed between wetter and drier. Blumenberg thinks that fossil remains of many animals still living in these regions today attest to the claim that conditions would not have been much different from the present.

Both chimpanzee species (*Pan troglodytes* and *Pan paniscus*) survive quite well on the savanna, and *H. erectus* migrated from Africa, traversing many variations in climactic conditions and eventually settling in a relatively wet forested ecosystem in Java, 1 million years ago. Blumenberg (1983) rightly concludes that environmental pressures were therefore not critical to the appearance of new biological adaptations within the evolving hominid lineage. Hominids, like many other primates, were equipped with the behavioural flexibility to track changing environments and persist. The environment did not 'select' for these traits, but rather accommodated any change in phenotype that appeared in our hominid ancestors.

2.9   Summary

Some, like Chomsky (1988), believe that adaptationist stories offer little explanatory value, and simply displace the problem of how initial mutations arise. Likewise, Bowers (2006) suggests that scholars should stop wondering about selective pressures and concentrate more on what kind of mutations might be involved in the formation of novel phenotypes. I have highlighted the shortcomings of selectionist theories and argue that we should follow the advice of Pigliucci and Kaplan (2006) that we "need the detective more than the statistician".

Darwin (1872) summed up his case that species have been modified "during a long course of descent" with a touching admission that he may have "underrated the frequency and value [of] variations which seem to us in our ignorance to arise spontaneously". In answer to his critics, he also thought that he had been quite conspicuous in his view that "natural selection has been the main but not the exclusive means of modification". Early in *The Origin of Species*, Darwin acknowledges the work by Weismann, who pointed out that the "nature of the organism" is the most important factor when considering how evolution can produce variation, over and above the external conditions of life.

Quartz (2003) bemoans the fact that much of evolutionary psychology is still caught in the grip of viewing the mind as a collection of modules that are separately heritable, and that evolutionary change is primarily seen in terms of changes in the gene frequency that determines these modules. In his view, failing to take account of developmental processes and constraints has resulted in a distorted view of the evolution of human cognitive architecture. He rightly complains that much of evolutionary psychology ignores development and that it is in fact mostly "adevelopmental" in approach. The changes that take place during the 'developmental program' of the brain produce systematic changes throughout the brain. Quartz believes that this offers strong evidence against the massive modularity thesis, and the independent evolutionary account of our cognitive traits.

A further problem with the selectionist account for the evolution of modules linked to behaviour comes from developmental considerations. For Quartz, it simply does not make sense to divide the neocortex into a collection of autonomous modules that could determine human traits. The restricted phases of neurogenesis could not possibly allow for independent traits to be under their own selective pressures. Stotz and Griffiths (2003) are also troubled with the adaptationist tradition, which, although it recognizes that phenotypic variation is the main requirement of evolution, its emergence is rarely addressed. The developmental systems that give rise to variation should enter into the equation before reference to inheritance in the Mendelian tradition. They challenge evolutionary psychology to not only specify domain-specific and content rich modules, but also to define the module developmentally. Any theory of evolutionary emergence should look to developmental structures and their relation to modularity. Stotz and Griffiths underscore the point that an understanding of how the mind grows is critical to determining how it might have evolved.

In the next chapter, I present a different take on the evolution of species. This non-adaptationist approach concentrates more on the interrelations between ontogenetic development and evolution, rather than forming theories of evolution under an adaptationist framework.

3. Saltationism and Evolutionary Development

3.1 Introduction

In this chapter, I explore some of the recent work in the field of evolutionary developmental biology (evodevo) that has important insights for how we might explain and account for saltational events. Specifically, I highlight the important role that development plays in producing organisms that conform to the 'normal' features that define its species, but, more interestingly, its role for those organisms that deviate from their regular or 'expected' developmental trajectory. These deviations often lead to speciation events. I also contrast the evolutionary developmental approach with neo-Darwinian, gradual, adaptationist theory for how the evolution of species proceeds.

Taken as an alternative approach to Darwin's theory of gradual change, saltationism[1] takes a more 'internalist', and non-adaptational view toward how organisms evolve. The role of development has a long history in biology, and was certainly popular in Darwin's day. Haeckel, a contemporary and great admirer of Darwin's work states that:

> All evolution, all development of organic individuals is in reality [in Warheit] epigenesis, i.e., an activity of life that rests essentially on the progress of

---

[1] Characterized as "cloud-cuckooism" by Dawkins (1991).

generation [Zeugung], growth and differentiation,
a transformation of similar [gleichartige] parts to
dissimilar parts, and an actual genesis of new individuals from non-individualised materials. In all of
these processes, which may express themselves as
development [Hinaufbildung], transformation and
involution, as progressive or regressive metamorphosis, the theory of epigenesis remains as the leading
principle [leitende Grundgesetz] (Haeckel, 1866).

The saltationist approach to evolution places much importance
on the mutation 'events' that led to the unique traits that define
species. A simple but laudable observation is that

knowing how something originated often is the best
clue to how it works (Deacon, 1997).

In a similar vein, Carroll believes that

just as for any work of human creation, we so much
better understand how complex things have come to
be – cars, computers, spacecraft – when we understand how they are made, and how each new model
is different from its predecessors (Carroll, 2006).

I agree with Marcus (2006) that Darwin's most important point
was that evolution proceeds by "descent with modification"[2].
Darwin also placed much importance on natural selection for the
evolution of species, but I argue that this offers little value for
explaining the evolution of novelty, and that it is of secondary
importance.

To be fair, Darwin also recognised the relevance of mechanisms
of development. He noticed that certain structures in organisms

---

[2]Although it must be recognized that this was not strictly Darwin's
revelation. Several scholars before him, including the French zoologist
Leclerc, the inventor of the term *biology*, Lamarck, and even Darwin's
grandfather, Erasmus, had clearly outlined their theories that most of life's
diversity had arisen from a common ancestor (Moore and Moore, 2006).

appear to grow in a correlated manner. His work with pigeons revealed that those with short beaks had small feet, and those with long beaks had large feet, a phenomenon he put down to "the mysterious laws of correlation" Darwin (1872). Here, Darwin had the key insights for mechanisms that only now are being discovered by developmental biology, where organisms appear to develop within a highly conserved but interrelated program of 'sets' of structures[3].

It has long been noted that the fossil record fails to show evidence of slow, generational change, or a trend toward complexity within species. Instead, we find sudden speciation events followed by long periods of stasis. For example, we have a sequence known as the Hamilton invertebrates, which comprises an entire fauna that appears suddenly in the fossil record around 380 million years ago. This group of organisms persisted with "great monotony" for 6 million years, then abruptly disappeared (Eldredge, 1996). As the environment changed, the fauna tracked that change, rather than changing their form to adapt to the changing environment. Most of the fossil record reveals a similar history.

Darwin grappled with this problem, because he believed it was most likely the main obstacle in the path of formulating his theory of evolution by gradual change.

> As according to the theory of natural selection an interminable number of intermediate forms must have existed, linking together all the species in each group by gradations as fine as are our existing varieties, it may be asked: Why do we not see these linking forms all around us? Why are not all organic

---

[3]Bowers (2006) has shown that the spine, pelvis, and limbs, as well as the jaws, appear to develop as a set. Interestingly, early hominids show modifications of all of these anatomical regions, indicating that they most likely arose due to a single deviation from the developmental 'program' of their pongid ancestor (e.g., Bowers, 2006; Gilbert et al., 1996; Hallgrimsson et al., 2007; Marcus, 2006).

beings blended together in an inextricable chaos?
Darwin (1872)

A classic Darwinian approach to the evolution of organisms
expects gradual, accumulative change, which transforms an
organism into a different species, usually assumed to be a fitter
or better adapted species to its environment. Schwartz and
Maresca (2006) suggest that the adherence to a gradual model
of phenotypic evolution may have actually been reinforced by
molecular systematics, which is founded on a "molecular clock"
model. This model is based on the assumption that mutations
in DNA molecules will happen gradually and continually. It is
commonly thought that minor point mutations will gradually
accrue and gradually transform an organism until it changes
into a recognizable new species. Most of the genome comprises
a vast amount of repetitive DNA that has resulted from sudden
replicatory events, and has had profound evolutionary effects.
The major changes in organisms, including humans, should
be understood as changes in the molecular structure of genes
that regulate development and yield structure, rather than
minor change in genes that produce minor variation *within*
populations.

By the 1970s, Gilbert et al. (1996) recognize that the 'gene-
centered' approach to evolutionary change was seriously challenged. Microevolutionary change, due to variations in gene
frequency, did not seem enough to be able to produce new
species. Adaptations may have allowed for "the survival of the
fittest", but not "the arrival of the fittest". The new challenge
to the received view of how evolution builds different species
focused on macroevolution, homology, and the morphogenetic
field. Hall (2002) points out that examples, like the macroevolutionary event that transformed the reptilian lower jaw into
the mammalian middle ear ossicles, heralded the epigenetic
approach to cells. The emergence of molecular biology has
allowed for the study of heterochronic changes during development, leading to functionally different phenotypes. Macroevolution was postulated as an alternative view to gradualism. The

adaptationist program was beginning to be challenged by both paleontologists and evolutionary biologists. Homology once again became the central concept. Mechanisms such as heterochrony were posited to demonstrate how macroevolutionary novelty can rapidly evolve (Gilbert et al., 1996).

Shifts in timing of development produces evolutionary changes in features already present in ancestors (Gould, 2000). The discovery of a sequence of stable genes (the homeobox) have allowed molecular biology to find many conserved homologies, including the *Pax-6* gene involved in specifying the transcription factor for development of the eye in both insects and vertebrates. Many other Hox genes have been found to be involved in the building of structures like limbs and pumping organs, to name just a few.

Gilbert et al. (1996) do not deny that genes are important for evolution and development, however it should be recognised that genes build cells, but gene 'products' need to interact to create morphogenetic fields, in order to build viable organs. Hlusko (2004) is confident that developmental genetics will be able to decipher genetic mechanisms that are responsible for minor phenotypic variation. Research will also be able to discover the higher level genes that control changes in developmental rate or timing during ontogeny. Small changes in these genetic mechanisms during development may produce major changes in developmental 'sets'. Higher level changes in the developmental genes, for example those that produce growth factors, can produce major variations in the ways animals develop. Gilbert et al. (1996) point out that if, for example, a growth factor stayed active for just one more cell division at a certain point in time during development, then an organ would be greatly enlarged. Natural selection could then promote or filter out this mutation.

The 'developmental-genetic toolkit' is an important 'player' in the development and evolution of novel phenotypes, however Newman and Bhat (2008) have drilled down to a different level of developmental processes. They have introduced the concept of 'dynamical patterning modules' (DPMs)[4] to show how physical processes involving chemical and mechanical interactions within and between cells constitute a 'pattern language'. This 'language' can explain the causal factors for the development of many (or all) multi-cellular organisms from the beginnings of life on Earth. They have shown that widely divergent body plans and organ forms could have emerged by exploiting the physics of "an extensive morphospace" of cellular interaction without reference to major changes in the 'toolkit' genes, or the need to invoke natural selection as a 'causal' factor[5].

Jablonka and Raz (2009) also stress the need to incorporate the different mechanisms of epigenetic inheritance within a "shared evolutionary framework". Epigenetics is the study of the processes that allow for both developmental plasticity and canalization. Epigenetic inheritance occurs when phenotypic variation occurs without any discernible changes in DNA base sequences. Their example is a case in point; an individual's kidney stem cells and skin stem cells share identical DNA sequences, but variations in developmental stimuli lead to different cell phenotypes.

Along similar lines, Palmer (2004) has found that genetic assimilation can occur where a novel phenotype arises *before* the genotype and may be a common mode of evolution. He points to the many organisms that show some form of asymmetry, and there are many plants and animals that develop as sinistral or dextral forms in equal distribution, regardless of the phenotype of their parents. A gene may be expressed in different tissues or at a different developmental stage, which, according to Palmer, implies different molecular mechanisms.

---

[4]Newman and Bhat (2008) describe DPMs as units consisting of one or more products of the 'toolkit' genes.

[5]Newman and Müller (2005) reserve a role for natural selection as a stabilizing force on functionally useful morphological outcomes.

3.2 Developmental Systems

Minugh-Purvis and McNamara (2002) highlight the advances in molecular biology that have led to the emergence of the discipline of evolutionary developmental biology. Variations in developmental rates and timing operate at levels from molecules to organ systems. Regulator genes, or Hox genes, operate at a different level from the typical variations that arise from the processes of inheritance by Mendelian genetics[6].

We know that there is no direct correlation between the number of genes that an organism has to the relative complexity of its phenotype. For example, the nematodes have upward of 21,000 structural genes compared with humans who have around 25,000 (Schwartz and Maresca, 2006). The difference between phenotypic outcomes has little to do with these structural (protein building) genes, but lies in the presence of regulatory genes. The location and timing of their expression during embryogenesis, or later developmental stages, is the result of transcription factors, which have increased threefold from a fruit fly to the human genome. Differential control of gene transcription is facilitated by gene-specific transcription factors.

A simple change in molecular signaling can change the developmental trajectory of an organism in a major way. Raff (2000) has shown that experimentally varying the resources available for the growth of a particular body part usually affects the size and/or shape of other parts. He finds that developing systems appear to have genetically discrete modules that interact epigenetically during growth. These modular interactions can become dissociated in timing, leading to changed patterns of development, but nevertheless producing a viable organism. As noted in the introduction, a concentration of regulatory genes located on the human chromosome 2 are involved in the devel-

---

[6]Although, as Maresca and Schwartz (2006) point out, Mendelism emphasizes discontinuous variation and is compatible with saltational evolution due to the introduction of novel features.

opment of the vertebral column, pelvis, limbs, hands and feet, as well as the jaw.

West-Eberhard (2005) recalls an interesting case of "Slijper's Bipedal Goat". Slijper, in 1946, had dissected a goat that had walked (or hopped) bipedally due to a developmental anomaly affecting its front legs, which were too short for normal quadrupedal locomotion. This goat had developed human-like gait characteristics. The induced move to bipedal behaviour caused changes in bone, muscle mass, tendon length, and extensive changes in pelvic and thorax shape. This suite of *correlated* changes is thought to be the result of epigenetic modifications of the phenotype, with little or no genetic change.

It is noted by (Raff, 2000) that early development often "evolves freely", allowing for highly divergent ontogenies to emerge among closely related species. For example, most extant crustaceans share a common larval form that has been highly conserved for over half a billion years! This initial common form evolves into the many thousands of species with quite different adult body forms. New developmental pathways can produce marked evolutionary novelties. Therefore, Raff believes that "evolution cannot be understood without understanding the evolution of development, and how the process of development itself biases or constrains evolution".

From an exaptational perspective, Marcus (2006) makes the point that novelty is usually a variation on a pre-existing theme, and almost nothing is without precedent. Much of evolutionary novelty is the result of a change in the normal developmental trajectory of the species. He points to the many skeletal parts, like the vertebrae of the spine, which consist of serially repeated subcomponents. Along similar lines, the four chambers of the mammalian heart, which resemble one another, have clearly evolved from the three-chambered heart found in amphibians, which in turn evolved with modification from the two-chambered heart of the fish (Marcus, 2006). The forelimbs in vertebrates – arms, flippers, and wings – are another example of an inherited scheme of descent with modification. In line with this scheme, he believes that current cognitive modules should be understood as products of evolutionary changes in ancestral cognitive modules.

Most modifications in animal bodies have resulted from minor changes in developmental patterns, and some of these changes have allowed for a radically different form of locomotion. For example, bats have an unexceptional mammalian body except for the forelimbs, which have elongated finger joints that act as struts for the support of membrane forming a wing (Raff, 1996). This forelimb emerges from a perfectly ordinary mammalian limb bud, which is part of the ancient pattern of limb structure that we find in the first Tetrapods. In the following chapter, we see how the changes in anatomy of the first hominids that led to bipedalism appear to have involved only some minor changes (or even a single mutation) in the developmental pathway of the spine and pelvis, together with the fore and hind limbs.

Many evolutionary theories, mostly in the Darwinian gradualist tradition, have focused on the comparison of adult morphologies, but Hall (2002) believes that it is very important to look at deviations from the ancestral form that may occur at any time during an organism's developmental trajectory. Similarly, Raff (1996) believes that too much evolutionary theory focuses primarily on how adult morphology may have evolved, rather than looking at the many transformations that have taken place in earlier life-stages of organisms leading to novel phenotypes.

An interesting discussion comes from Trut et al. (2009), who refer to experiments with the domestication of foxes. The domestication of foxes has been achieved by selection for tameability and has opened up the possibilities for ascribing the causative role of the underlying molecular genetic mechanisms involved in these changes. They found that certain behavioral traits (tameness) correlate with the morphological changes associated with delayed development. The wide skull, shortened

snout, floppy ears, and curly tail are all juvenile traits and are
associated with a delay in developmental rates observed early
in embryonic development. It is believed that the physiological changes related to domestication may result from changes
in a very small number of brain genes with many regulatory
downstream effects. Here we have another example of the
importance of considering epigenetic modifications and their
impact on evolutionary theory.

3.2.1   Homeobox – Hox Genes and Development

Hox genes, together with other transcription factors, are mostly
responsible for the regulation of patterns of development (morphogenesis). Maresca and Schwartz (2006) have stressed the
importance of the increase in complexity of transcription factors
during the evolution of life. Yeasts have only 300 transcription
factors, while humans have around 3000, an increase that they
believe is likely to be responsible for much of the genomic diversity and organismal complexity over time. Their model of
evolution takes into account the importance of the structure
and mechanisms of DNA replication, and also effective repair
mechanisms. Accordingly, most evolution of complex shapes
and structures has arisen due to the shifting of when and where
various developmental control genes are turned on or off during
development. A slight mutation in the DNA sequence of a Hox
gene can alter the timing of morphogenesis, and thereby effect a
change in identity of an embryonic region. This developmental
approach to the evolution of novel anatomical traits precludes
the theory that anatomical structures have evolved by the gradual accretion of minute changes, in the Darwinian adaptationist
tradition. Rather, genome sequencing provides evidence that
evolution has proceeded from large-scale duplication, deletion,
or merging events, or even whole genome duplication.

Lovejoy et al. (1999) are critical of theories of natural selection that have most often posited the heritability of 'atomized'

traits, due to the false presumption that particulate inheritance underlies morphological traits. They also suggest that just because a trait can be separately defined and analyzed, this does not mean that it has a direct analogy of a unique and isolated genetic 'blueprint' for that trait. Their work has shown that the hominid pelvis, with all of its components, is an important example of rapid and dramatic morphological change. The entire hominid pelvic field and the femur has been simultaneously altered by a slight change in signalling molecules, which assign positional information to cells that determine their dimensions during embryonic development. An interesting although speculative theory has been put forward by Bowers (2006), who points to the human chromosome 2 which has formed from the fusion of the pongid chromosomes 12 and 13. This fusion may have had extraordinary impact on the developmental outcomes for the first hominids. A concentration of regulatory genes involved in the development of the vertebral column, pelvis, limbs, hands and feet, as well as the jaw, are located on the human chromosome 2. This chromosomal fusion is likely to have produced the crucial mutation that created a cascade of developmental changes in the first hominids. This fusion can explain the sudden emergence of all of these altered traits, which seem to arise as a developmental 'set'.

Genes code protein sequences made up of the amino acids that are synthesized elsewhere in the cell and build their respective proteins in most living organisms, from nematodes to humans. Hox genes, or regulatory genes, have been found to encode transcription factors that regulate the growth processes and often delimit relative body regions (Bogin, 1999). A change in these transcription factors can have a dramatic impact during the early stages of the life-history of an organism. A simple quantitative change in one part of development can have flow-on effects in other regions. Hox genes appear to have been highly conserved throughout most of evolutionary history on Earth. However, regulatory elements can duplicate or rearrange causing their expression to shift in time, or be expressed in a different site than normal, allowing for the formation of novel connections

(Raff, 1996). Each organism exhibits a developmental path that is highly interactive. Hall (2002) has warned against the positing of an evolutionary change at one level alone, which is often the approach taken in classical adaptationist theory. Limb and digit reduction in tetrapod evolution can be directly linked with the involvement of homeobox genes Hoxa-11, Hoxa-12 and Hoxa-13. A mutation in the Hox-13 gene instigates abrupt changes in the number and shape of digits in the forelimbs or hindlimbs. The HoxD gene complex, located on the chromosome 2 in humans, is active in the formation of the lower vertebrae, pelvis and posterior limbs (Bowers, 2006). We must also remember that the *Bauplan* of the limb involves both genetic and epigenetic processes. The developing limb is the result of many molecular and biomechanical regulatory processes that are "beyond the strictly genetic" (Newman and Müller, 2005).

Adherents of Darwinian adaptationism often refer to the evolution of the eye as being a classic case of an advantageous trait being selected and enhanced over time. Once again, it can be shown that the development of the vertebrate eye is under the control of a small number of developmental controlling genes. Schwartz (1999) has shown that a small molecular change can even produce an effect as profound as having, or not having, an eye.

Growth hormones, under the control of developmental genes, affect the developmental pathways of all mammals. The most obvious effects of growth hormones at critical developmental stages can be seen with the variations in dog breeds, which we know belong to the same genetic species. These highly variable phenotypes are due to simple variations in level of a particular growth hormone (IGF-1), which has local effects in various parts of the body, especially in limb and skull proportions (Fondon and Garner, 2004).

3.2.2 Paedomorphism

A particular form of heterochrony produces paedomorphosis, which is the retention of infant or juvenile characters into the adult stage of development. It can be shown that paedomorphic processes have been important in much of evolution, and these processes are of special interest when it comes to hominid evolution. A simple change during development can produce a radically different phenotype in a single generation.

The most often quoted extant example of paedomorphism is the change in the developmental pathway of the axolotl. The axolotl remains in its juvenile state, due to the failure of the hypothalamus to produce thyroxine at the right developmental phase. Failure of a single molecule to initiate anatomical metamorphosis thus causes it to retain a paedomorphic, juvenile lifestyle, although it can nevertheless produce viable offspring. The juvenilized appearance is due to cellular insensitivity to a key hormone (McKinney, 1998; Raff, 1996). Crockford (2002) has found that the thyroid hormone is strongly implicated in control mechanisms for developmental rates and timing during embryonic development in most organisms, including humans, in the early embryonic stage as well as postnataly. The thyroid gland influences the transcription of many genes including those that synthesize growth hormones. A small change in this mechanism can influence cell migration and differentiation during both embryonic and postnatal growth in brain development. Other influences can be seen in adrenal gland function, hair production, and skin and hair pigmentation, leading to slight variation within species. However, growth hormones are also able to cause major phenotypic changes, which may actually isolate a species from its founder population.

Schwartz (1999) refers to the often ignored puzzle of how the first vertebrates evolved. They appear suddenly in the paleontological record: where are the transitions? The most elegant explanation is underpinned by the processes of heterochrony.

One invertebrate species, the tunicate, passes through a free-swimming larval stage where its form resembles that of the first chordates. It has a cartilage rod and a major nerve trunk along its back. Upon attaining the adult form and sexual maturity, it changes dramatically into a sessile, barnacle like invertebrate. Schwartz (1999) and Gould (2000) have theorized that an ancestral chordate could easily have retained its larval stage, producing the first viable vertebrate. This 'sudden appearance' of a new species is simply a paedomorphic form of its immediate ancestor. In this case, there *are* no missing links, and there is no gradual change leading from one species to another.

Sauropods are a group of dinosaurs that had a quadrupedal gait. However, Reisz et al. (2005) have discovered articulated embryos of a group of prosauropod (*Massospondylus carinatus*) that indicate how quadrupedal dinosaurs may have evolved through paedomorphosis. The adults of *Massospondylus* were bipedal, but were quadrupedal as infants. It appears that the quadrupedal posture of sauropods may have evolved through the retention in the adult of the early limb proportions of the young. Rather than following the 'normal' ontogenetic pathway, leading to bipedalism, this group of dinosaurs retained the long forelimbs of their hatchling form.

Groves (1989) has found homology of some of the cetacean body forms with the features of a pig embryo. Due to a heterochronic mutation, the body form appears to be arrested at the time of the formation of limb buds, and the body retains a hairless, fusiform shape with no demarcation between body and tail. There are no external appendages, such as ears, and the animal retains a glabrous skin. Tooth development reaches only a simple tooth bud stage in baleen whales and then partly resorb[7]. In addition, the skeleton has not differentiated, and, as a consequence, there are hardly any hindlimbs and the *manus* have not reached pentadactyl form. Many whales retain a vestigial small pelvis as a result of this interruption in growth. This

---

[7]See section 4.5 for details of the incredible amount of plasticity of mammalian teeth during development.

change in developmental trajectory has an interesting consequence in that it occurs at a time when the brain is growing fastest, and therefore produces the relatively large brain seen in most cetaceans. During embryonic development of cetaceans, the nostrils start off in the 'normal' mammalian place, but then move to the top of the head to form one or two blow holes. Recent molecular DNA evidence also supports the close relationship between whales and the hippopotamus (Boisserie et al., 2005). It seems that cetaceans (the order to which the whales belong) are closely related to the hippopotamus and the pig, both of which belong with the artiodactyls (even-toed placental mammals with hooves). The hippopotamus is hairless like the whale, nurses its infant under water, and also communicates by underwater sound, just like a whale. At face value, it would be very difficult to find a selectionist account of how this radical mutation would have enhanced the fitness of this mammal. Nevertheless, we can see how a major mutation, due to an arrest of fetal development, may have produced an animal that seems to have spawned several highly successful cetacean species, including over one hundred species of whales. Eldredge (1996) points out that whales appeared suddenly in the fossil record around 55 million years ago. This radically new mammalian phenotype fortuitously found a niche, the ocean, that would support its ecological needs. The animal did not 'adapt' to an oceanic environment, but rather an existing environment 'accommodated' a new species that was already 'adapted' to life in the ocean.

It seems possible, or likely as I argue, that a change in developmental trajectory in our *H. erectus* ancestor provided the crucial and sudden mutation that lead to the emergence of *H. sapiens* anatomy. The fossil record, although sparse when it comes to finds of infants and juveniles, attests to the claim that modern human anatomy arose in an 'ultra-saltational' event with the retention of the shape of the infant *H. erectus* skull and face into the adult stage. *H. sapiens* do not follow the developmental pathway of *H. erectus*, which lead to the growth of large brow ridges, a receding forehead, and a prognathic jaw.

Instead, adult humans retain the rounded skull with more or less bulbous forehead, flat face, small teeth, and the position of the skull immediately atop the spinal column – all traits found within the fossil record of several young *H. erectus* species. I present this case in more detail in the following chapter.

3.3   Summary

In this chapter, I have argued that evodevo yields a more plausible picture for how novelty arises. Changes in developmental timing have been shown to have played a major part in the evolution of many species, including hominids. Neoteny, involving the extension of infant growth rates into sub-adulthood, appears to have had a major influence on the evolution of not only the first hominids, but also modern humans. Bipedal locomotion afforded by the retention of a juvenile shaped pelvis, reduced prognathism leading to a change in dentition with the late development of the canines, and the position of the skull directly atop the spine, are all features that can be explained by the arrest of developmental processes in our pongid/hominid ancestor (Gould, 2000; Montagu, 1989).

The sudden appearance of anatomically modern humans, around 120,000 years ago, has not been, and cannot be, explained by traditional Darwinian theory, whereby species are thought to emerge gradually, with features having been 'honed' by natural selection. Recent fossils finds of infant and juvenile hominids are shedding light on how *H. sapiens* emerged in just one crucial step. Modern humans have not followed the developmental pathway of their hominid ancestors, but rather, have retained the infant morphology of their immediate *H. erectus* ancestor. Even the emergence of our enlarged brain can be explained by a critical change in developmental timing, which allows time for the massive proliferation of glial cells, the growth of dendrites, and the growth and myelinization of axons (Gould, 2000). This massive increase in white matter, supporting and enhancing the neurons, has lead to major advances in memory retention and cognition.

The remaining chapters of the book expand on the line of argument developed in this brief chapter, and do so in the context of examining, respectively, paleontological, archaeological, genetic and linguistic evidence. In backing saltationism over gradualism in this theoretical struggle over the specifics of evolution, I am well aware of the problem of the underdetermination of theory by evidence. I can see no better way of arguing in favour of saltationism than showing that it provides the best fit and the most plausible explanation of all that evidence.

4. Paleontological Evidence

4.1 Introduction

In the previous chapter, I highlighted the importance of how simple changes in the developmental pathways of organisms can give rise to major evolutionary novelty. In this chapter, I outline the relevance of this approach to the evolution of the hominid clade. I will show that modern human anatomy did not evolve through gradual change. An additional claim will be made that the brain did not evolve gradually by incremental change, neither in size nor in complexity.

Here, Darwin's claim that evolution proceeds by 'descent with modification' takes precedence over his idea that evolution happens by gradual change through adaptation and is 'driven' by external selection pressures. Although humans are undoubtedly descendants of an ancestral hominid line, I will argue, following Tattersall and Schwartz (1998), that we need not assume that *we* are the culmination of a long line of gradual and generational change in populations over a vast amount of time under the action of natural selection, as often claimed by many scholars (Buss, 2005; Carroll, 2006; Cosmides and Tooby, 1994; Dawkins, 1991; Deaner and van Schaik, 2001; Milton, 2006; Moore and Moore, 2006; Pinker and Bloom, 1990, for example).

A 'textbook'[1] Darwinian explanation for how humans evolved

---

[1] Browse through any textbook on offer to students of human evolution, and you can usually find some sort of banal commentary on how or why hominids became bipedal, usually without any reference to paleontological or archaeological evidence.

by natural selection can be found in the following passage by a
leading primatologist.

> Darwin (as usual) got the narrative of human evo-
> lution about right. His version was that we left our
> long-term home in the African forests to begin our
> lonesome journey into the savannah, which then
> selected us to walk on our hind legs. The challenges
> of savannah carnivores and competition within our
> own species led to brain growth. Big brains endowed
> us with the gift of group morality, which let us co-
> operate against carnivores but also make war on
> one another, using weapons and tools. From there,
> group support and transmitted material culture led
> on to the triumphs-or complexities-of civilization
> (Jolly, 1999).

My saltationist position is the antithesis of this approach to
the evolution of both hominid anatomy and human cognition.
Hominids did not 'choose' to leave their long-term home in the
forests and then 'adapt' to the savannah by 'becoming' bipedal.
One of the earliest hominids, *Ardipithecus ramidus*, considered
by many to be a likely candidate for an ancestor of humans,
was clearly bipedal, but also retained limbs that allowed for
agility within the forest trees (Pi et al., 1997). Moreover, there
are just as many carnivores in the forest as on the savannah,
so it is not apparent that early hominids faced more challenges
of a predatory nature. Also, why would we assume that early
hominids were more competitive within their own species on
the savannah than in the forest? Why would hominids grow
large brains due to these presumed additional pressures while
other primates did not? An additional problem arises with
Jolly's picture of the evolutionary 'progress' of hominids when
we consider the fact that hominid brain size did not actually
increase from that of chimpanzees, when body size is taken into
consideration.

As for warlike behaviour, we have no evidence of any 'con-
structed' weapons for any time during hominid evolution up

until historical times, and these weapons are only associated with late modern humans. Hominids may have engaged in warlike behaviour in the same sense that extant chimpanzees (*Pan troglodytes*) commonly 'interact', but the archaeological record does not offer any clues to support the notion that their conflicts were any different from other animals. We also have to ask where Jolly finds the evidence for tools, group support and transmitted material culture. Apart from stone tools that arrive in the landscape five million years *after* the evolution of hominids, this evidence simply does not exist.

The evolution of hominid species, eventuating in the appearance of modern humans, has followed a path of sudden mutations during the developmental phase of the immediate ancestral species, with the latest crucial mutation causing the emergence of *H. sapiens*. The picture is one of step-wise evolution followed by long periods of stasis, rather than gradual adaptational evolution. Up until the sudden emergence of *H. sapiens*, there is a lack of evidence that would point to any important cognitive advance over that of extant pongids.

It has long been recognized that significant changes can evolve during any developmental stage, be it embryonic, fetal or juvenile. For example, Montagu (1989) views the evolutionary pattern of hominids as a process of 'growing young'. I will show that the hominid fossil record reveals some of these apparent developmental changes that have led to the retention of infant or juvenile traits, starting with the earliest hominid ancestor and leading up to the emergence of modern humans. Schwartz (1999) points out that most animals have a juvenile morphology quite different from their adult form, especially in the proportions of the head, eyes and face. This is especially apparent in apes, where the eyes get proportionally smaller, and the flat face develops into a snout as they pass through the juvenile stage. During the developmental program, the globular skull gradually transforms into an entirely different shape, and large brow ridges emerge. Humans experience a redirection of onto-genetic trajectory where the adult skull and face retain similar proportions to their own juvenile form. A study by Naef (1926) of the differences between a juvenile and an adult chimpanzee, was a catalyst for the idea that humans may have emerged due to the neotenous retention of our immediate ancestral infant form. The photographs on the front cover of this book, taken by Herbert Lang, were apparently much published in the early 1900's and appeared in Naef's paper.

There are several major complexes that we equate with the evolution of humans,

1. Bipedalism.

2. Gracility of the face and jaws with reduced canines.

3. Hands with precision grip.

4. Modified vocal tract.

5. Large and Complex brain.

In this chapter, I present compelling evidence that points to the developmental processes involved in the emergence of these human traits. Heterochronic changes, especially neoteny, will be shown to be heavily influential in the evolutionary modifications of our Great Ape ancestor right up to the emergence of *H. sapiens sensu stricto* only around 120,000 years ago. The fossil record will be garnered to support the theory that these anatomical changes were neither the result of adaptations to external forces, nor did they arise by gradual, generational change. Climate change[2], changes in social structure, sexual selection pressures, or any other external 'causes', did not 'drive' evolutionary change in hominids.

My approach will necessarily be grounded in the processes of developmental change that have caused novel phenotypes

---

[2]Caveat; rapid environmental change can induce significant mutational events due to changes in the expression of regular amounts of stress proteins responsible for protein folding, and thereby increase the mutation rate and genetic change during development (Maresca and Schwartz, 2006).

to arise. Hominids, like many other organisms throughout evolutionary time, emerged due to mutations that appear to have 'arrested' or 'extended' a particular developmental stage[3] of their ancestral species. Evidence from the fossil record shows that our hominid ancestors had a radically different life-history from *H. sapiens* (Tardieu, 1998).

4.2 Heterochrony and Hominid Evolution

It is apparent that most of the changes in an organism's lifetime occur during growth and development. An interruption of the 'normal' trajectory of development can have major effects both in morphology and behavioural traits. Comparative primate morphology reveals that the differences in form between humans and apes are produced by alterations in the rate and duration of development, and are mostly quantitative in nature (Gould, 1977).

We know, for example, that all Great Apes have tails during fetal stages of development, and that this tail is resorbed due to a heterochronic change in the expression of homeobox genes (Schwartz, 1999). The resulting 'mutation' has quite dramatically changed the behavioural abilities of apes from their monkey cousins. We shall see later in this chapter how heterochronic changes have shaped our pelvis, spinal column, cranium, face, jaws and teeth, and lastly, but most importantly, our brain.

4.2.1 Neoteny and Hominid Evolution

Neoteny, involving the extension of infant growth rates into sub-adulthood, has been heavily implicated in the evolution of

---

[3]The term stage is used loosely while recognizing that development involves a cascade of complex epigenetic and mechanical interactions.

hominid anatomy. The pattern of change due to these neotenic processes should however not be seen as gradual, but rather as a staged effect. The first major mutation produced a bipedal, paedomorphic primate, which appears to have retained the same anatomy, more or less, for over five million years.

In the preceding chapter, we explored the developmental pathways of a wide range of organisms, which, throughout evolutionary history, have been influenced by sudden deviations from their 'normal' developmental program. We saw that even the very first vertebrate was most likely the result of retention of the larval stage of its invertebrate ancestor (Schwartz, 1999). We also saw how closely cetaceans (whales, dolphins, etc.) resemble the pig embryo at the point of development of the limb buds, where the body takes a fusiform shape (Groves, 1989). This evolutionary novelty seems to have arisen 'suddenly' without the necessity for millions of years of gradual, Darwinian evolution.

Paedomorphism has been part of the hominoid developmental *Bauplan* for many millions of years before the emergence of hominids. Bipedal locomotion, reduced prognathism leading to reduced canines and sometimes extreme microdontia, and precision grip in the hands, arose long before the first hominids were found in Africa. This type of evidence challenges many of the speculative stories that have been abundant throughout the twentieth century. Stories like 'we needed to stand upright to free our hands for tool use', or 'our canines reduced because tools took over as weapons', are not tenable. For example, Pinker (1997) thinks that our delicate faces emerged because tools and technology took over from teeth. This is an especially strange proposition due to the fact that stone tools appear in the archaeological record many millions of years after bipedalism emerged in the first hominids, and our delicate faces emerged only 120,000 years ago.

The retention of juvenile morphology is evident within the hominid clade, and here we will explore some of the evolutionary 'sudden' appearances of our ancestral forms that are due, in part, to this process. Montagu (1989) has observed that the juvenile skulls of the first hominids[4] found in Africa at that time (australopithecines) resemble modern humans more closely than the adult australopithecine form. This observation also applies to the later species (*Homo erectus*), and is especially apparent, when we note the shape of the skull of the infant *H. erectus*. The Neandertal evolutionary path diverged from our hominid ancestral line over 500,000 years ago, but even in their ontogeny we find 'hints' of human morphology in the juvenile form. Analysis of Krapina 1, a juvenile Neandertal of approximately 7 years of age, has been put forward as evidence of continuity of form between Neandertals and modern humans, but its slightly modern appearance is suggested to be, in part, due to its young age (Minugh-Purvis and Radovčić, 2000).

Neotenous processes appear to have preserved the shape of the infant *H. erectus*[5] skull and face into the adult stage in modern humans (Arsuaga et al., 1999b). Modern humans do not develop the large brow ridges, receding forehead, and prognathic jaw that developed in the adult *H. erectus*. Rather, adult humans retain the rounded skull with more or less bulbous forehead, flat face, small teeth, and the position of the skull immediately atop the spinal column. It seems that any of our hominid ancestors could have been 'predisposed' to a saltational mutation leading to the neotenous phenotypic traits of modern humans.

---

[4] As more fossils are found in Africa, an ongoing controversy ensues over the diversity of hominid species that date to 6-7 millions years ago. Schwartz (1999) points out that although these early finds are more apelike than later australopithecines, they are classified as hominids due to hints of bipedalism based on the forward position on the skull base of the articulation with the vertebral column.

[5] I refer to *H. erectus* as our immediate ancestor, but treat it as a grade due to the fact that the variations (e.g., *H. ergaster*, *H. heidelbergensis*, *H. antecessor*) appear to have shared a similar morphology as infants (Antón, 2002, 2004; Arsuaga et al., 1999b; Baab, 2008; Gould, 2000; Nishimura et al., 2006) and consequently any one of these hominids could have been our forebear.

4.2.2   No Trends

The first split from our Great Ape ancestor, as argued later in this chapter, appears to have been merely a bipedal ape with little change in life-path development from extant chimpanzees. A developmental stasis lasted for up to five million years until the emergence of *H. erectus*, offering little credence in support for theories of gradual evolution of hominids *leading* to modern humans.

A recurring theme in this chapter is the argument that hominids did not evolve as a 'trend' toward modern human anatomy. We did not 'gradually' become bipedal, or taller, or reduce the size of our brow ridges[6] or our teeth. The shape of our skull, face, and oral cavity did not gradually change from robust to gracile. More importantly, we did not 'gradually' grow larger and more complex brains. The picture, as exposed by the fossil record, is one of a series of 'sudden' mutations, followed by a mosaic[7] of hybrid forms and then stasis, lasting for millions of years

---

[6]Due to the fact that neo-Darwinians *expect* to find a neat progression of hominids leading from ape to modern human, they generally classify fossils by looking at morphological features that have "slightly more archaic features" (Wong, 2003), but offer little evidence that has anything to do with any *advance* in terms of being advantageously selected. Brow ridges, for example, have often been used as a feature that defines how 'primitive' a hominid is without any regard to the fact that we do not see a gradual reduction in brow ridge size over time within hominid anatomy. In fact, some Neandertal and *H. erectus* specimens had larger brow ridges than extant chimpanzees. Brunet et al. (2002) have suggested that brow ridges may have been under strong sexual selection, which is a strange statement considering modern humans males have virtually lost them. A developmental explanation seems more appropriate than a selectionist account. The disappearance of brow ridges can be easily explained by the paedomorphic retention of the infant facial morphology of our immediate *H. erectus* ancestor (Gould, 2000; Arsuaga et al., 1999b; Nishimura et al., 2006).

[7]We could leave open the possibility that this diversity of hominid form could have resulted through traditional Darwinian adaptational evolution. However, it is difficult to find any plausible accounts of gradual evolution of any of the traits that define early hominids or the later australopithecines, leading to greater fitness.

after the first split from our Great Ape ancestor[8], and then over one million years, possibly up to two million years, for our presumed immediate ancestor, *H. erectus*.

Much controversy surrounds the number of hominid ancestors that may have existed since the split from our Great Ape ancestor. Eckhardt (2000) points out that the number of species of fossil hominids is naturally based on the perception of morphological differences, but he suggests that the difference between hominid populations may actually be about the same as the difference between dogs and wolves. Gene exchange would have been possible between these variable hominid taxa so that they may represent just one diverse species. We should also keep in mind, as Ankel-Simons (2000) points out, that chimpanzees that belong to the same species have a high degree of facial feature variation. This standpoint is supported by Eckhardt's work among other primate groups, like macaques and baboons, where phenostructure and zygostructure do not coincide. He argues that many inferences based on morphological differences may misrepresent evolutionary relationships. Rather, polymorphism within a single species, as we see in other primate taxa, seems to have been the situation. More specifically, he argues that this approach to hominid diversity precludes the notion of a trend from more to less 'primitive' morphology. He points out that although the difference between extant chimpanzee and human genome may be around 1000 genes, there is no reason for assuming that the rate of mutation and accrual of mutations has been uniform. The apparent reduction of the chimpanzee 48 chromosomes to 46 chromosomes in humans, he suggests, would have generated major genetic and phenotypic variation.

Recent analysis of cranial shape variation in our putative ancestor, *H. erectus*, by Baab (2008), points to a single species with very limited variation, despite its extensive geographic range

---

[8]As evidence accrues for defining *Homo floresiensis* as an australopithecine, we may be able to extend this genera's existence for up to seven million years.

and having to deal with novel climatic conditions. Despite some minor intraspecific variation, which is to be expected in such a widely dispersed species, Baab observes that the neurocranial shape was conserved for the entire existence of this species. She concludes that *H. erectus* did not undergo speciation from the time of its emergence 1.8 million years ago, right up until its demise between 300,000 and 100,000 years ago. Supporting Baab's conclusions, we have the most recent fossils of *H. erectus* from Dmanisi, Georgia (Lordkipanidze et al., 2013). Five individuals, who belong to the same early population of *H. erectus*, have widely different morphological features and brain size. If these five individuals had been found in different regions in Africa, it is likely that they would have been incorrectly classified as different species.

The underlying theme that I want to sustain in the next few sections is for the sudden appearance of many of the traits that appeared in the evolutionary path of our hominid ancestors. Once these traits emerged in our ancestors, there followed large periods of stasis of form lasting for millions of years. These sudden appearances can be accounted for by changes during the 'normal' developmental pathways of their ancestors. The apparent variations in phenotype can be accounted for by the effects of hybridization, as seen in not only other primate species, but in many other animal species as well.

Under Darwinian gradualist evolutionary theory, novelty is assumed to arise after the accumulation of minor mutations, which gradually transform the organism over many thousands or even millions of years as a result of natural selection. In contrast to this approach, we can see that novel phenotypic traits can emerge suddenly during development of the organism. It is important to note that many phenotypic changes emerge as the product of complex, multifactorial biological mechanisms (Pagni and Baccetti, 1993).

Many mutations arise in the 'recessive state' (i.e. inactive) and have no immediate effect on its bearer. However, inbreeding inevitably increases the rate of spread of recessives and increases the probability of producing homozygotes for this state. Accordingly, new species can appear to arise seemingly instantaneously without invoking geographic or any other isolating mechanisms (Maresca and Schwartz, 2006).

4.3 Bipedalism

One of the quintessential traits that defines us is that we walk on two feet. Falk (1992) even goes as far as thinking that bipedalism is the key to understanding human brain evolution. In this section I challenge this assumption. Humans are obligate bipedal primates due to changes in developmental processes that depart from the 'normal' pattern of ontogeny of primates. I will show that bipedalism is not a trait unique to humans, so we need not evoke a 'special case' for hominids in order to derive the 'reason' for its emergence. Nor do we have any reason to conflate bipedalism with any improvements in brain size or architecture, leading to more complexity of behaviour. We are bipedal because the anatomy of our spinal column and pelvis more or less force us to be. I restate my claim that until the emergence of *H. sapiens* we have every reason to believe that hominids were simply bipedal apes.

There have been many theories put forward in order to 'explain' why bipedalism arose. Under an adaptationist perspective, it is assumed that bipedalism must have conferred some sort of advantage to the first hominids. For example, it is often uncritically assumed that the first hominids were 'forced' out of the forest environment and onto the savannah due to climate change. This change in habitat is deemed to be the external force that 'drove' just some primates to adopt bipedalism in order to 'adapt' to the open plains. However, the sudden emergence of bipedalism in our Great Ape ancestor may, in fact, have had a negative impact on these primates. For some of these bipeds, bipedalism may have incurred the loss of protection of their arboreal refuge due to the loss of agility. Pi et al. (1997) have pointed out that australopithecines would most likely have retained their arboreal lifestyle, nesting like today's chimpanzees, in order to protect themselves from predators. There is no evidence for the use or control of fire associated with these early hominids, so they did not have this deterrent to predators. The earliest possible evidence for the control of fire dates to 180,000 B.P. at Kalambo Falls, Zambia, some many millions of years after the first hominids emerged. It would simply have been a matter of survival for our hominid ancestors to maintain a refuge in the rainforest canopy. Many conflicting theories have been advanced regarding the advantages or otherwise of bipedalism. Some scholars have argued that bipedalism may not have been a favourable mutation, due to the costly energetics associated with this form of locomotion. Stern (2000) has found that *Australopithecus afarensis* maintained a type of bent-hipped, bent-knee gait during bipedal locomotion, that was energetically costly. In contrast to Stern's analysis, Leonard and Robertson (1997) have suggested that the energetic savings of bipedalism in the first hominids would have given them a major boost in survivability, despite being tied to a forest environment. They believe that the energy savings of early bipedality in australopithecines would have meant that these hominids could have greatly expanded their daily foraging ranges over and above that of other apes. As the environment became more open and with food resources covering a greater area, these hominids would have found this locomotor shift an important factor for energy saving, perhaps enhancing the survival prospects for the species.

Whatever the effect that bipedalism had on the fitness of hominids, we can categorically state that its emergence was due to a sudden transformation followed by many millions of years of more-or-less stasis of form. In chapter 2, I challenged some of the 'explanations' put forward as Darwinian selection theories for the emergence of bipedalism. I conclude that we have no cause to assign any special status to bipedalism. Following from this, we also have little reason to believe that the emergence of bipedalism correlated with any advance in cognitive ability for our hominid ancestors. My aim is to focus on *how* bipedalism arose in our ancestors, not *why*.

4.3.1 Primate Bipedalism

In this section, I make the point that many of the phenotypic traits that we associate with hominids were already present in the developmental 'options' of other primates. There are no Great Ape fossils between 14 and 7 million years ago in Africa. Most fossils of apes have been found in central Europe, Greece, Turkey, South Asia and China (Begun, 2006). One ape, *Dryopithecus*, which appears to have evolved around 14 million years ago, is considered by Begun to be the ancestor of the African Apes. Another related Eurasian ape, *Oreopithecus*, which evolved around 7 million years ago, shows some very interesting traits. Some of theses that we associate with hominids, like bipedalism, precision grip, and smaller canines, were already present in one of our close Great Ape ancestors.

A species of this genera, *Oreopithecus bambolii*, evolved in the Miocene Mediterranean island of Sardinia. The lumbar region of this ape indicates lumbar lordosis just like the human condition. Interestingly, the anatomy of the pelvis resembles that of the first hominids and the early *Homo* clade. The femur and a tripod-like foot also correlate with bipedal activity, although Köhler and Moyà-Solà (1997) suggest that this ape may have been a bipedal shuffler rather than having the full functionality of modern human bipedalism. Alba et al. (2001) believe that the noticeable reduction in canine teeth of *Oreopithecus bambolii* was the result of its paedomorphic cranial morphology. This ape not only walked bipedally, but also had a human-like hand, which was capable of a precision grip. The hand of *Oreopithecus* had a firm grasping capability that is not found in other apes, although it is also found in the first hominids (Moyà-Solà et al., 1999).

Many species of apes flourished in the lush environments of Eurasia, but climate change around 9 million years ago caused most of them to go extinct. Two lineages persevered by tracking the subtropical environments in which they had evolved within Eurasia (Begun, 2006). One lineage moved into the African tropics and the other into Southeast Asia. One of these ancestral apes from Europe evolved into the first African hominid.

The first find of an australopithecine fossil was a 3-4 year old juvenile named by its discoverers the Taung 'child'. Its features, including the pattern of convolutions on the endocast, details of milk teeth and position of the spinal column, indicate that its early maturation pattern was the same as apes (Falk, 1992). The discoverer of this fossil, Raymond Dart, emphasized its 'human-like' attributes, such as the rounded forehead and position on the spinal column, indicating possible bipedalism. If we look at the developmental process of primates, we can easily see how retention of the juvenile physiology could have enforced bipedalism on the first hominid.

All primates are born with the long axis of the head approximately perpendicular to the vertebral column. At birth, the head sits directly on top of the spinal column in humans, monkeys, and apes. During juvenile development in apes however, this angle changes as the foramen magnum moves posteriorly, causing the head and the vertebral axis to move toward the same plane (Gould, 1977). This shift in position of the head in relation to the spinal column in the adult chimpanzee gives the impression of a 'stooped' posture, as can be seen in the image of an adult chimpanzee on this book's cover. Humans retain the fetal primate condition throughout their ontogeny, with the head positioned directly atop the spinal column, aiding an upright posture. Here we have just one possible explanation for the emergence of bipedalism. The first hominids may have simply retained the more-or-less erect posture of their infant condition throughout their development and into adulthood. In the next section, I outline other credible theories for how bipedalism may have arisen in our hominid ancestors.

4.3.2 Hominid Bipedalism

The evolutionary history of bipedalism reveals a path of sudden
mutation followed by a long period of stasis of form. Hominids
did not become bipedal under a scheme of selection and adaptation. The earliest fossil hominids dating to around 6-7 million
years ago have been found in Africa. These hominids are believed to have survived in Africa right up until around 1.2 million
years ago (Lahr and Foley, 2004)[9]. The defining features of
australopithecines are bipedalism, small stature (1-1.5 metres
tall), and reduced sized canines. The australopithecines show
a mosaic of ape-like and human-like features including tooth
size and shape, face shape, brow ridges, brain case size, femur
shape, pelvic size and shape, and various configurations of foot
bones that may mean full bipedalism, or for some, retention of
the ability to retain an arboreal, climbing lifestyle. Ackermann
et al. (2006) believe that much of the variation of this nature
can be accounted for by hybridization, which often leads to the
kind of novel phenotypes that we find in hominids.

Australopithecines retained long ape-like upper limbs and other
ape-like characteristics of the lower limbs, suitable for climbing
(Stern, 2000). This ape-like anatomy of limb proportion was
retained by hominids for about 4 million years, right up to and
including the emergence of *Homo habilis*, around 2.6 million
years ago. It is interesting to note that even *H. habilis* retained
a divergent big toe similar to australopithecine, which may have
meant that this species was still an efficient climber. Recent
work by Haeusler and McHenry (2007) shows that the upper
limbs of some *H. habilis* were similar to human proportions,
but the lower limbs were more ape-like. More importantly,
the oldest australopithecine, *A. afarensis* had more human like
limb proportions than the later species, *A. africanus* and *H.
habilis*. Even the limited samples of *H. habilis* fossils show

---

[9]Although, a member of the australopithecine clade may have existed
on the island of Flores in Indonesia up until around 18,000 years ago (see
section 4.7.3).

variation in proportions. Haeusler and McHenry are sure that this hominid lineage does not represent an evolutionary pattern of unidirectional and progressive change in limb proportions. The idea that there were any trends 'toward' human anatomy is not supported by the available fossil evidence.

Bipedalism is enabled in humans by several anatomical distinctions, namely the shape and position of the spinal column, the shape of the pelvis, and the anatomy of the femur, knee and foot. So far, we have two possible mechanisms that may have altered the usual development pathway of the Great Ape locomotor system. Recent analysis of the Chimpanzee genome has thrown light on the possible genetic cause of the mutation that changed the anatomy of the hominid locomotor system. Pongids have 48 chromosomes while humans have 46 chromosomes. Our chromosome 2 has resulted from a fusion of the pongid chromosome 12 and 13. The HoxD gene complex is located on the human chromosome 2, and Bowers (2006) believes that the mutation that produced the fusion of the pongid 12 and 13 chromosome may have also altered the HoxD sequences that are active in the formation of the vertebral column, pelvis and limbs. This reduction in the number of chromosomes is believed by Bowers to be the quintessential mutation that caused bipedality. It may also have triggered a speciation event due to reproductive isolation. The important point for Bowers is that we have a concise mechanism that could have produced this sudden emergence of evolutionary novelty. Interestingly, Christiansen and Chater (2008) use the emergence of bipedality to support their argument against the sudden emergence of novel traits where "no single gene creates an entirely new means of locomotion from scratch". Bowers has shown how a simple mutation *can* actually produce saltatory change and emergent novelty – in this case bipedalism.

Following the australopithecines (although the two species seemed to have overlapped in Africa for 600,000 years), we find the first hominid classified as *Homo*, which we know as *Homo erectus*. This species was bipedal, had a much larger body size, and a brain of between 750 cc to 1000 cc (Walker and Shipman, 1996). Some *H. erectus* left the African continent and migrated to Europe and Asia. With the evolution of *H. erectus* we find a quite radical transition to more human-like post-cranial anatomy, mostly with the length of the femur, which in humans is diagnostic of an adolescent growth spurt (Tardieu, 1998). The transition from *Australopithecus* to *Homo* appears to have involved heterochronic processes, which according to Tardieu, were most likely due to changes in hormonal control of growth processes initiated by the brain, or more precisely, the hypothalamus.

H. erectus was named for the first fossil hominid found in Java in 1892 by Dubois, a Dutch physician working in Sumatra. The fossil femur indicated that this hominid stood upright, hence the name *erectus*, and Dubois determined from this find that this hominid was ancestral to modern humans. Apart from a few differences, like a more rounded shaft and a variation in the position of the attachment of muscles, this femur was overwhelmingly human-like. Along with the femur, Dubois found a calvaria (skullcap), which resembled a chimp-like skull, although the brain size was ascertained to be about two-thirds of that of a human. Schwartz (1999) notes that it became clear to Dubois that our ancestors had become human-like first from the waist down. Hominid bipedalism had obviously evolved before a human-like brain.

According to Smith and Tompkins (1995), *H. erectus* had a life history quite different from both chimpanzees and modern humans. It seems that *H. erectus* did not experience the adolescent growth spurt that modern humans go through at puberty. For example, an African male *H. erectus* specimen had reached the height of a modern human adult male by 7 years old, but its dental and bone pattern of growth match that of a chimpanzee. The *H. erectus* brain size was unremarkable, in fact, only the size to be expected for a primate of that body size. Just as with the first hominids, we have little reason to assume that full bipedalism was in any way linked to any significant advance in cognitive ability. A survey of the archaeological record in the next chapter supports the conclusion that *H. erectus* made little, if any, advance in cognitive behaviour over and above its australopithecine predecessor.

The following sections will detail some of the evidence for how minor genetic changes during development have reshaped the typical chimpanzee-like locomotor system into the bipedality of hominids. A recurring theme in these sections will be the importance that neotenous processes have played in the emergence of many of the complexes that have enabled hominid bipedalism.

Primate Hind Limbs

Limb and digit reduction in tetrapod evolution can be directly linked with the involvement of homeobox genes Hox-11, Hox-12 and Hox-13 (Raff, 1996). The developmental process is similar for all Tetrapods, and a mutation in the Hox-13 controlling gene can cause abrupt changes in the number and shape of digits in the forelimbs or hind limbs. A simple mutation in the homeobox genes controlling limb development can cause the sudden appearance of radically altered shapes and structures, as we see in a bat's forelimb or a horse's hoof.

During human development, the human hand and foot emerge as paddle-shaped ends on the limbs, and the regions between the developing fingers and toes are formed due to cell death controlled by specific developmental genes. Gould (1977) points out that during primate embryonic development, at the stage when the digits have just separated on the plate-like hands and feet, there is no sign of rotation of the thumb or the great toe. At seven weeks gestation, all primate fetuses have differentiated fingers, but the digits of the feet are still adducted (fixed) (Montagu, 1989). The great toe can rotate, as in apes, or remain adducted, as in humans. Gould is convinced that the unrotated great toe in hominids was the result of a simple quantitative alteration in developmental rates. Our unrotated big toe already existed as a 'transient' stage of 'normal' primate embryonic development, but its paedomorphic retention has provided hominids with a useful functionality supporting plantigrade bipedal locomotion.

Primate Pelvis

The main difference between the ape and hominid pelvic bone is the short and stocky pattern of the latter, although many of our earlier ancestors, including australopithecines, *H. erectus*, and the Neandertals, retained an ape-like straight and slender iliac crest (Marchal, 2000). Analysis of pelvic bone of apes, hominids, and modern humans reveal two periods of stasis of form of the pelvic bone, once *Australopithecus* had separated from the Ape form. The *Australopithecus* pelvis remained unchanged until *Homo* arose, and Marchal thinks that this would account for a locomotor variation between these two ancestors.

A second period of stasis existed between *H. erectus* and the emergence of the modern pelvis of *H. sapiens*. Most importantly, Marchal's analysis reveals that the emergence of the *H. sapiens* pelvis was most likely the result of the retention of a paedomorphic morphology. His analysis is supported by others in the field. As noted previously, the fossil remains of the bipedal ape *Oreopithecus bambolii* reveal that paedomorphism can alter the shape of the 'normal' ape pelvis, resulting in a hominid-like structure (Köhler and Moyà-Solà, 1997). The cancellous bone architecture of the hip bone is intimately related to the mechanics of locomotion, and this pattern can be observed in *Oreopithecus* as well as humans and fossil hominids. This ape, which evolved many millions of years before the emergence of australopithecine, also maintained lumber lordosis like that of humans. It seems that paedomorphic developmental processes changed the anatomy of the pelvis thereby contributing to obligate bipedalism.

We can therefore challenge the assumptions of theories that posit external selection pressures for the emergence of bipedalism. Bogin (1999) states that "bipedalism is known to have changed the shape of the human pelvis from the basic ape-like shape". Here he is positing a change in 'behaviour' as the 'cause' of physiological change. The more plausible account comes from developmental biology. The paedomorphic change in the anatomy of the pelvis changed the locomotive behaviour of our hominid ancestors. We also can see that this important part of our anatomy did not emerge in a Darwinian adaptational fashion, but rather by heterochronic saltation (paedomorphism). After the paedomorphic pelvis emerged, we find a vast period of stasis of form numbering millions of years.

Spinal Column

Analysis of the vertebral column has exposed some surprising facts, which support arguments against a neat 'progression' of the evolution of the anatomy that supports bipedalism, the vestibular system supporting balance, and breathing control supporting human language. We see a picture of mosaic evolution rather than any 'trend' toward adaptation of hominid anatomy for bipedalism, fine motor control and breathing.

Walker and Shipman (1996) describe the most complete skeleton of *H. erectus*, the Nariokotome boy, which was found in the Turkana region of Africa. The *H. erectus* fossil had 6 lumbar vertebrae, a configuration that is found in the first australopithecines as well as with the presumed ancestor of the Great Apes and the hominid line, Proconsul. This configuration had not changed in over 20 million years! Modern humans have 12 thoracic vertebrae (one for each pair of ribs) and 5 lumbar vertebrae, one less than our ancestor, *H. erectus*.

Once again, we find developmental processes implicated in the evolution of hominid vertebrae. Different morphological characters seen in anterior and posterior vertebrae are determined by regulator Hox genes. Raff (1996) has found that a simple knockout of the *HoxA-5* gene can cause the transformation of one type of vertebrae into another. The result of this experiment caused the identity of an anterior vertebra to take on a similar shape and size of a posterior vertebra.

Although *H. erectus* had a different configuration of the spinal column, it nevertheless had a vestibular system like a modern human, indicating an ability for fine balance and agility. Walker and Shipman (1996) make the interesting point that the presumed first tool maker, *H. habilis*, had a vestibular system more like a gibbon, and appears to have been less bipedal than its predecessor, australopithicene. Once again, we see that there were no gradual trends 'toward' bipedalism.

The features of the spinal column in all hominid species show a mosaic of developmental changes and cannot be viewed as adaptations of increasing complexity supporting bipedalism, motor control, or, as argued in the next section, language.

Spinal Column and Implications for Speech

All primate spinal cords contain both gray and white matter, with the gray matter consisting of nerve cell bodies and the white matter making up the nerve fibres, which are covered with a fatty, myelinated sheath. Compared to chimpanzees, humans have a dramatically increased amount of gray matter in the spinal cord, which takes up most of the room in the newly enlarged spinal canal. MacLarnon and Hewitt (2004) have analyzed the differences between the gray and white matter, which makes up the soft tissue in the spinal cord. This enlargement is directly related to the innervation of motor functions controlling the thoracic and abdominal muscles implicated in the fine control over breathing. This novel and more complex configuration has given modern humans a much greater control over posture, coordinated bipedalism, fine motor control with the hands, and most importantly, control of breathing. Walker and Shipman (1996) point out that this last enhancement has important implications for the evolution of language.

An important anatomical difference that we find in *H. erectus* is the size and shape of the individual vertebrae. The vertebral body surface is more ape-like than that of a human. The canal that carries the spinal cord is much narrower, in fact only half the size of that of a human. Humans need the fine control over their abdominal muscles to control breathing as they speak. The fact that *H. erectus* had a spinal column more like an ape than a human, casts doubt on many of the abilities that are unquestionably attributed to this species, namely coordinated running and walking, fine motor control over the hands for making stone tools, and breathing control necessary for spoken language.

MacLarnon and Hewitt (2004) point out how important fine neurological control of the respiratory muscles is for human speech. The production of speech in humans, or vocalizations in other primates, involves the lungs acting as bellows to push air through the larynx. This pulse of air causes the vocal folds to vibrate, thus producing sound. Humans are able to finely modulate this sound due to the configuration of our long pharynx, or upper throat, tongue, and lips. The pharynx is much longer than in other primates due to the descent of the larynx to a position lower in the neck. Our fine control of breathing allows for the wide variety of output, from punctuated bursts to smooth continuous sounds. The rapid manipulation of the upper respiratory tracts allows humans to produce phonemes in sequences around 10 times faster than non-human primates. Of course, it is a stretch of the use of the term phoneme to describe primate vocalizations. Most non-human primates can only produce one unmodulated sound per breath movement, which is limited by the maximum rate of cycles of inhalation and exhalation. Apes tend to vocalize on both inhalation and exhalation as can been seen in the pant/hoot of chimpanzees. Humans however, usually only produce speech on the exhalatory phase, which may last from 2 to 6 seconds, but may extend to

12 seconds.

The fine control that humans have over their breathing does not result from any basic design changes or size of the lungs. All mammals have about the same lung volume, scaled to body size, and their lungs share the same elastic properties. Humans use subglottal air pressure to emphasize syllables or phonemes at the required moment. The meaning of a phrase can change dramatically due to the different emphasis on a word, as in the example with the statement that "pop took his socks off" (MacLarnon and Hewitt, 2004). Non-human primates however, do not have this capability. Their vocalizations are produced with the same intensity, which fades as the lungs deflate. Human speech relies on the rapid feedback and coordination of the upper respiratory tract and also the respiratory muscles that control the fine movements of the lungs to produce all of the various linguistic demands. All of this feedback and integration involves highly sophisticated neural control, enabled by our newly evolved spinal column.

MacLarnon and Hewitt (2004) suggest that the capacity for speech was a recent phenomenon in hominid evolution and may have been made possible due this greater control over breathing. Neandertals seem to have developed larger thoracic vertebrae, which may have enabled greater control over breathing. Although they may have had the capacity for fine control over the production of speech sounds, evidence from the configuration of the basicranium in these hominids indicates that the larynx was still high in the throat. Raff (1996) reminds us that the upper respiratory tract of human newborns resembles that of other primates so there is little doubt that our hominid ancestors had the same configuration in infancy. It is only at the age of two years in humans that the larynx descends in the neck. It appears that the descent of the larynx was not part of the developmental pathway for ancestral hominids, casting doubt on their ability to maintain spoken language.

It is probable that much of the anatomical support for spoken language arose recently. This argues strongly against any notion that these traits were 'fine tuned' to support language in a gradual, adaptational fashion. Together with the complete lack of evidence for symbolic 'behaviour' in our hominid ancestors, it seems reasonable to conclude that speech as we know it arose only with the emergence of anatomically modern humans, around 120,000 years ago.

4.4   Hands with Precision Grip

Adaptation is a secondary utilization of architectural novelties, but as Finlay et al. (2001) note, those who put adaptation forward as the architect of evolution can "with enough ingenuity" rationalize anything as some sort of adaptation. For example, the oldest intentionally modified stone tools known as the Oldowan assemblage are dated to around 2.5 Mya and are believed to have been produced by an early *Homo* species. The really large finds are dated from 2 to 1.5 Mya coinciding with the emergence of *H. erectus*. Based on the evidence that chimpanzees and orangutans regularly use tools for digging, cutting, extracting nuts, and termite fishing, Panger et al. (2002) reasonably surmise that early hominids would have had the capacity for the same sort of tool use. However, due to the fact that *Australopithecus africanus* dated to 2.4-3 Mya, seemed to display human-like characteristics of the hand, they propose that their hands were slowly "adapted" for manipulation, despite the complete absence of intentionally modified stone tools associated with them. The manual dexterity needed to make stone tools came *before* the actual arrival of stone tools.

A more parsimonious explanation for the evolution of the human-like hand comes from Hlusko (2004), who points out that many anatomical traits come as developmental sets. The modifications in the pelvic region, the feet and the hands of *Australopithecus* arose due to a mutation that affected the developmental, functionally interrelated set of traits. The saltatory event that produced bipedality seems also to have enabled the development of a hand with shorter fingers. Hlusko has shown that shorter fingers are likely to be a pleiotropic effect of shorter toes. The development of fore- and hindlimbs share the same patterning mechanisms, so a developmental change in one parameter can effect limbs or even digits in both. The fore- and hindlimb have evolve independently in some animals, but a general trend of correlation seems to hold for hominids. Bowers (2006) believes that the immediate developmental consequence of bipedality may be that areas in the sensory and motor homunculi of the brain, that were previously devoted to the grasping ability of the foot, were co-opted for manual dexterity. Rather than adaptive function "driving" architecture, we see a classic case of what Finlay et al. (2001) have justifiably argued to be secondary utilization of architectural novelty.

4.5　Primate Crania - Face, Jaws and Teeth

One of the most dramatic changes that take place between the neonate and adult ape can be found in the size and shape of the crania. *Dryopithecus*, which evolved around 14 million years ago in Eurasia, is considered to be the ancestor of the African apes because this species carries the ground plan representing many of the unique aspects of adult cranial form of the African Great Apes (Begun, 2006). The skull of Chimpanzees and Gorillas changes dramatically from that of the infant as development progresses to the adult stage.

The most interesting point here is that the *Dryopithecus* skull shows a morphology similar to that of the extant chimpanzee *juvenile*, not the adult. Both *Dryopithecus* and the bipedal *Oreopithecus* retained a paedomorphic cranium with a rounded braincase, indicating that neotenous processes were at work during their evolution. As noted above, these same neotenous processes appear to be responsible also for bipedalism in *Oreopithecus bambolii*.

Oreopithecus bambolii had a marked reduction in canine teeth, which was the result of a paedomorphic cranial morphology (Alba et al., 2001). This paedomorphic ape exhibited a reduction of facial prognathism, which resulted in a lack of space to accommodate adult dentition. The canines, which are the last teeth to develop in apes, even after the second and third molars, are believed to have reduced in size due to the lack of developmental potential (Schwartz, 1999). Interestingly, a similar morphology is found in extant *Pan paniscus* (Bonobo), which also is believed to be a paedomorphic ape. We find that many of the features that define *H. sapiens* cranial features, including the shape of the cranium, the flat face, the small jaws and the configuration and size of the teeth, to be the result of similar neotenous processes in our evolution.

McCollum (1999) has made an extensive study of the hierarchical development of the primate face and has determined that the unique australopithecine face can be interpreted in the context of growth remodeling and displacement during ontogeny. A 'developmental cascade', involving remodeling of the entire hominid face, is due to bone deposition and resorption and has resulted as developmental by-products of dental size and proportions. The huge molars together with the reduced incisors and canines required the remodeling of the jaws, hard palate, and the nasal and oral cavities. It seems quite puzzling, if we were to speak in Darwinian selectionist terms, why these hominid ancestors lost the dagger-like canine teeth of their Great Ape ancestor. Considering fitness, long and sharp canine teeth are extremely useful in battle, and for killing and tearing the flesh from prey. Some Darwinian selectionist accounts suggest that the loss of these large canines in our hominid ancestors may have been the catalyst for 'inventing' stone tools. Of course we know this cannot have been the case, as stone tools do not appear in the archaeological record for at least 3 millions years *after* the emergence of the first hominids with reduced size canine teeth.

When we come to our immediate ancestor, *H. erectus*, we find increased flattening of frontal contours of the skull with increasing developmental age (Antón, 2002). By adulthood, *H. erectus* skulls sloped almost directly backwards from their massive brow ridges. The change from the infant shape of the skull to the shape of the adult skull is quite dramatic. This massive transformation is entirely unlike the developmental changes in modern humans (*H. sapiens*), who retain the rounded shape of the infant vault together with a flat face, right into adulthood. In fact, some of the first fossil humans found in Africa had extremely neotenous skulls with a ratio of cranium to face of five to one (see section 4.2.1). Interestingly, the unique human chin arises due to one part of the jaw being retarded in growth (Gould, 1977). The neotenous retardation of the jaw in comparison with other primates, during ontogenetic development, also produces a decrease in size of the dental arch and a more prominent nose. These features arose as an indirect consequence of the paedomorphic retention of orthognathy (flat face).

Of course, the retention of a juvenilized skull and face does not automatically equate with an advanced cognitive ability. Recent excavations of a 800,000 year old cave site in Atapuerca, Spain, have uncovered the skull of a juvenile hominid, which has the face of a totally modern human (Arsuaga et al., 1999b). Arsuaga, believes that this find most likely explains how modern humans emerged due to a developmental change that caused the retention of the juvenile morphology (neoteny) of an ancient species named *Homo antecessor*.

Kunzig (1997) believes that the modern human face of the juvenile of this 800,000 year old species raises some interesting questions relating to the nature of human evolution, both in 'technology' and behaviour. The tools found alongside *H. antecessor* were surprisingly archaic for their time, resembling the 2.6 million year old Oldowan tradition of the first tool makers in Africa. That is a period of nearly 2 million years without any advance in sophistication of a tool-kit. In addition, there is evidence of cannibalism of the young members of the species. Although recent human groups have been known to slaughter one another for ritual purposes, the victims of *H. antecessor* had been dismembered and butchered like any other animal. The remains were found along side those of other animal bones that were stripped of anything edible in the same way. Bermúdez de Castro et al. (1999) suggest that the site indicates gastronomic cannibalism rather than a survival strategy, and excludes any possibility of ritual intentionality. They question whether gastronomic cannibalism could ever be an evolutionarily successful strategy, especially when the victims were infants and young individuals. No other artifacts and certainly no evidence of any ritual behaviour are found with these remains.

Most importantly, these fossil remains indicate that certain modern human anatomical traits were already evident within the juvenile stage of an ancestral hominid species. They indicate that certain 'options' were available for inclusion in a new species

(*H. sapiens*) by alteration of life history parameters during development of our predecessor. This snapshot of a hominid population from 800,000 years ago also reminds us that evidence for modern human *anatomy*, at any stage of evolution within an ancestral species, does not entitle the assumption that modern human *cognition* or behaviour (including language) evolved in parallel with these anatomical changes.

Dentition

The eruption of dentition is one of the most defining traits of life stages in primates. Fully functioning dentition allows for a certain amount of independence of a juvenile from parental care, especially feeding responsibilities. Due to the fact that teeth fossilize extremely well, they are used as one of the main diagnostic tools for determining species categorization. Tooth morphology can identify a fossil species, and eruption patterns together with tooth wear can determine the age of the individual at death. This is particularly important in order to determine the life-history stages of our ancestral hominids. Patterns of enamel growth during development can also determine the health profile of the individual. This section will highlight the fact that human dentition did not evolve as a 'trend' from the 'primitive' to the 'modern', nor did dentition evolve in a Darwinian, gradual pattern.

The first australopithecines had ape-like, rectangular dental arcades. It was not until the evolution of the first homo species, *H. habilis*, millions of years later, that a more human, parabolic dental arcade, arose (Walker and Shipman, 1996). Dean et al. (2001) make the important point that "an understanding of developmental processes can provide powerful insights into the evolutionary history of adult morphologies". They have performed extensive analysis of the enamel formation trajectories for the teeth of a range of hominids from *H. habilis* through to *H. erectus* and have found that none of them fall within the samples from modern humans. Enamel thickness has been used to determine whether a hominid is more closely related to apes, which have quite thin enamel, or to humans, who have evolved a relatively thick enamel. All primates secrete an enamel matrix in a circadian manner, that is, as daily enamel increments. Scanning these layers with electron microscopes reveals the enamel growth rates of each tooth.

Analysis of our hominid ancestors shows that they had similar enamel thickness to modern humans, but it had developed along the same growth trajectory as African Apes. Humans (*H. sapiens*) form thick dental enamel along a different developmental process. These results have led Dean (2000) to recommend that at least all of the early *Homo* species, including *H. habilis* and *Homo rudolfensis*, be returned to *Australopithecus*. Early hominids dated between 1.8 and 3.7 Mya show the maturational profiles of modern Great Apes. Dean finds that even our presumed immediate ancestor, *H. erectus*, does not exhibit the same tooth development of modern humans. Our 18-20 year period of growth and development evolved recently, after around 17 millions years of an ape-like life history profile (Dean, 2006).

Interestingly enough, our so-called closest 'cousins', the Neandertals, had thinly enameled teeth (Tobias, 1998), a fact that challenges adaptationist 'stories' for the evolution of teeth as a trend toward becoming 'human-like'. We can clearly see that one of the key traits that is used to define modern humans is not a trait that has evolved in a gradual and incremental fashion.

Dean (2000) proposes that the most secure suggestion for the first shift that we see in enamel growth patterns is with modern humans and perhaps the larger brained *H. sapiens* by 100,000 years ago. The pattern of tooth development also has important implications for the maturation pattern of brain growth, as we find later in section 4.7. Enamel growth trajectories reveal a significant difference in maturation patterns between humans and our hominid ancestors, and this finding is supported by tooth eruption patterns.

Smith and Tompkins (1995) have determined that australopithecines erupted their dentition along the same trajectory as chimpanzees, but we need to account for the variation in size of the canines. From an evolutionary development viewpoint, we can easily see how this reduction in canine teeth could have emerged suddenly in the first hominids. This reduction in canine size is not a unique phenotypic trait of hominids. As noted above, small canines have been part of the *Bauplan* of other paedomorphic primates. We find a fossil hominoid species (*Ouranopithecus macedoniensis*), that flourished in the subtropical environment of Eurasia around 15 millions years ago, which exhibits canine reduction characteristic of early hominids and unlike Chimpanzees and Gorillas (Richmond, 1999). Interestingly, the canines of this species do not have honing facets, indicating that they were not used in antagonistic encounters, as with other predators, which use their canines as weapons. As we saw previously, paedomorphic cranial morphology leads to lessened facial prognathism, which in turn results in a smaller jaw and less space to accommodate the canine tooth, which is the last to erupt in apes (Alba et al., 2001).

Certain Hox genes are known to control the shape as well as the numbers of sets of teeth, varying to a great degree between different classes of animals. Ancestral species of fish had endless replaceable sets of deciduous teeth, while primates have only one set of 'milk' teeth. Tooth formation in hominids can be shown to follow directly from the influence of Hox genes. Even in modern humans, if a tooth bud does not have the developmental competence to reach a critical size threshold, then it will be resorbed or remain stunted. We can appreciate how the canine in hominids became stunted due this lack of developmental competence. We can also dismiss theories that argue for gradual reduction of the canine in hominids. Jablonski et al. (2002) for example, believe that there was a 'trend' toward a reduction of canine teeth, and a reduced sexual dimorphism,

due to less agonistic encounters in the earliest hominids. The
fossil evidence does show a reduction in size of the canines, but
this was an abrupt change, and not the end product of any
trend toward lessened use by our hominid ancestors.

After extensive analysis of the fossil record of the dentition of all
H. erectus fossils, Rightmire (1990) has found that these fossils
fall into resemblances with either the first australopithecines
or with *H. habilis*. Also, the same range of variation is found
within all *H. erectus* fossils, indicating that there was also no
'trend' toward a reduction in the size of the molars from early
H. erectus to late *H. erectus*.

H. sapiens have markedly reduced dentition compared with all
other hominids, but this represents a 'sudden appearance' rather
than the end product of an adaptive trend. *H. erectus* remains
covered Africa, China, Europe and South East Asia throughout
the Pleistocene, but Rightmire (1990) has concluded that this
species maintained a conservative morphology over this entire
time. *H. erectus* features did vary, but these variations were
not on a scale greater than those found within contemporary
modern human populations. These observations are backed up
by the latest fossil find of *H. erectus* individuals in Dmanisi,
Georgia. These five fossil individuals show that, what appears
on the surface to be remarkable morphological variation *within*
the same population, is not dissimilar from modern human
variation (Lordkipanidze et al., 2013). These variations can
also be found within geographically widespread groups, and
cannot be placed within a neat time-line from primitive to more
human-like. Crania from Africa and Java also resemble one
another in both size and overall appearance. Variations in the
size of both these crania and the jaws can be easily explained by
sexual dimorphism. Rightmire (1990) has found that *H. erectus*
teeth vary slightly, but reduced sized molars appear 'suddenly'
in the fossil record associated with *H. sapiens*, and are not the
'end product' of a gradual change.

4.6 Evolution of the Vocal Tract

The human vocal tract seems to have been 'designed' as a near perfect system for speech production. Due to this apparent 'design' of the modern human vocal tract, it is often assumed that it *must* have developed gradually over a period of millions of years.

Bower (1989) believes that this argument connects to a broader scientific dispute over whether evolutionary change typically proceeds incrementally or in bursts after long periods of stability. Many have argued (e.g., Ploog, 2002, Pinker and Bloom (1990), Pinker (1994)) that the human vocal tract has evolved gradually, having been fine tuned over perhaps millions of years, serving an ever increasing complexity in communication in the form of articulate language.

However, this approach assumes that language has evolved *for* communication. Arguing against this scenario, Chomsky (2005) points to imaging studies that lend support to the fact that areas of the brain dedicated to the "*function*" of human language "*structure*" appear to be independent of speech and sound. This suggests to Chomsky that the sensiormotor systems supporting language in communication are not a special adaptation *for* speech or sign, but rather have been co-opted or recruited by the language faculty. Piattelli-Palmarini (1989) finds that vocal tract physiology provides a mechanism of selection for a limited set of phonological parameters matching those available in modern-day languages, rather than having been honed to accommodate speech. Of course, having generally agreed that language seems to be a more or less innate capacity in humans, it raises the interesting question of why deaf babies babble with their hands. It seems more likely that language emerged as an internal reasoning tool, which appears to be expressed externally in whichever motor output system it can utilize.

When it comes to finding evolutionary precursors to the human vocal tract, we are confronted with a major challenge. Most of the soft tissues of the vocal tract (the pharynx, larynx, tongue and muscle attachments) do not fossilize. The only clues that we have concerning the anatomy of the vocal tract come from indirect parts of the hominid skeletal anatomy.

The Hyoid Bone

The hyoid is a small U-shaped bone, which anchors muscles connected to the jaw, larynx and tongue. The only fossilized hominid hyoid has been found with a 60,000 year old Neandertal skeleton from the Kebara cave site in northern Israel (Bower, 1989). The shape and size of this Neandertal hyoid is similar to that of a modern human. This find has led to speculation that Neandertals may have been capable of human-like speech.

Bower (1989), however, points out that the hyoid bone alone, whether in modern humans or apes, cannot predict the shape and position of the larynx. Lieberman (1992) agrees that the physiology of the hyoid bone cannot be used to diagnose speech ability as even the pig has a similar hyoid bone to that of humans. Bower concludes that we will never be able to use the hyoid bone as an indicator of how the vocal tract of our hominid ancestors was arranged. Consequently, it cannot be used as evidence for the question of whether Neandertals could speak. Noble and Davidson (1997) also believe that the size and shape of the hyoid is irrelevant to any discussions on whether language was possible, as the *nature* of the utterance is the point at issue, not its production involving the throat.

The Basicranium

The base of the skull (basicranium) is an important part of skeletal anatomy and can provide indirect evidence for the presumed position of the larynx (Bower, 1989; Nishimura et al., 2006). Its form can be used as diagnostic evidence to predict where the larynx would have been positioned within the neck of hominids. At birth, the contour of the base of the skull is relatively flat in both apes and humans. As the human infant grows, the skull base bends, the face remains short and flat, and the larynx descends. This change in morphology produces a long pharynx, which provides for a huge variety of sounds over and above those that are possible for other primates.

Comparative studies of mammals show that a relatively straight, unflexed, basicranium indicates that the larynx would sit high in the neck (Bower, 1989). The basic mammalian pattern locks the larynx into the nasopharynx, which is the air space at the back of the nasal cavity. This system allows for a direct connection from the nose to the lungs. Animals with this configuration can still breath and drink at the same time. Human infants have an elevated larynx like any other mammal, which allows them to still breath while suckling. When the larynx drops, by the time a human is one and a half years old, the pharynx, an air cavity that is part of the food pathway, is large enough to shape sounds generated by the vocal cords.

Hominids of 4 million years ago had flat, ape-like basicrania, indicating that the larynx would have been positioned high in the neck. Later hominids (*H.erectus*), according to Bower's research, had the basicranial flexing equivalent to that of a six year old human. In hominids living 400,000 years ago (*H. heidelbergensis*), the basicranium is flexed similar to that of a modern human, although the face of the adult was unretracted. As we see in the next section, the neotenous retention of the infant flat face in humans is a key developmental trait linked to the production of speech. Bowers thinks that the anatomy of the *H. heidelbergensis* basicranium may have been 'language ready' before the emergence of the modern human skull, but these hominids showed no sign of having reached any advance in cognitive development over that of other *H. erectus*.

Facial Flattening

Recent MRI studies by Nishimura et al. (2006) show that facial
flattening appears to be the major cause for the development
of the human supralyngeal vocal tract (SVT). The neotenous
evolution of the human face has produced facial flattening with
reduced prognathism and projection. *H. erectus* infants had flat
faces, but during development they increased in prognathism
thereby remodeling the SVT. The reduced growth of the palate
during infancy and the juvenile period in humans has built a
SVT where the horizontal oral and vertical pharyngeal cavities
are in quite different proportions to chimpanzees. This unique
configuration of the SVT allows humans to extensively modify the resonance properties that in turn modify the laryngeal
sounds. These laryngeal sounds form the complex sequential
phonemes of speech in a single short exhalation. Nishimura et al.
point out that facial flattening most likely evolved only secondarily, and not initially to support speech, but later facilitated
the evolution of spoken language.

The Hypoglossal Canal

The hypoglossal canal is one more part of the hominid anatomy
that has recently been the subject of debate on whether our
hominid ancestors were capable of speech. This canal provides
a passageway though bone for the cranial nerve bundle that
innervates the tongue. DeGusta (1999) has reported, after
extensive analysis of both extant primates and fossil hominids,
that all of these non-human hypoglossal canals fall absolutely
and relatively within the size range of modern humans. That
is, the hypoglossal canal of all australopithecines and modern
humans studied do not fall outside the size range for all African
apes.

DeGusta therefore concludes that the hypoglossal canal is defi-
nitely not a reliable indicator of speech capabilities. Although

the neural innervation of the tongue is important for human speech, the organization of information and motor output necessary for speech occurs in several regions of the cerebral cortex, and the evolution of these regions of the brain are probably more important (DeGusta, 1999).

4.7 Brain Growth and Size

In this section I show that the human brain did not emerge as the end product of gradual Darwinian evolution. Although we do not have fossil brains, we can estimate brain size, and also use endocasts to determine, albeit roughly, certain brain architectural regions. The following section will highlight the fact that for most of hominid brain evolution we have a picture of brain size stasis when we take into consideration body size. This scenario contrasts with a gradualist picture where it is assumed that the brain gradually increased in size (e.g., Milo and Quiatt, 1993 who quote a size increase of 50 cc per 100,000 years for *H. erectus*), and state that this 'gradual' change would equate with a gradual increase in cognitive ability.

4.7.1 The Australopithecines

Hominid brains four million years ago were approximately 450 cc, whereas modern humans have a brain size of around 1400 cc. Falk (1992) has studied the endocasts of most of the australopithecine collections and confirms that they clearly indicate an apelike cerebral cortex. Even the most recent australopithecines, who lived up until around 1.2 million years ago, had ape-like cerebral cortices.

As noted earlier, although early hominid brains were sometimes slightly larger than those of their Great Ape ancestors, this slight enlargement could easily be accounted for by a slightly

larger body size. These first hominids had brain sizes that did
not lay outside the 'normal' brain for body size of any other
primate. Australopithecine brains were the same size, or in
some cases, only slightly larger than extant chimpanzee brains
(Eldredge, 1996).

H. habilis, the presumed first hominid to make stone tools, has
been put forward by some as the first species to show a marked
increase in encephalization. However, Ankel-Simons (2000)
has shown that this species, along with the later *H. erectus*,
are quite different from *H. sapiens* in skull morphology. The
forehead expansion in *H. habilis* and *H. erectus* is not equated
with an expansion in the frontal lobe size, but was a necessary
requirement to accommodate the insertion of large muscles that
operate masticatory processes, as we find in Orangutans.

There are no stone tools associated with hominids before the
emergence of *H. habilis*. The archaeological record is completely
silent for this vast amount of time. Australopithecines did
not make stone tools, nor did they leave any other signs of
technology or cultural practice. Hence we have little reason to
surmise that they had made any cognitive advance over their
chimpanzee-like ancestor.

4.7.2 *H. erectus* Brains

In this section, I aim to show that even with the emergence of
our purported immediate ancestor, *H. erectus*, the same picture
persists with the brain, that is, one of sudden emergence due
to an increase in body size, followed by a long period of stasis.
This, of course, is not the pattern of brain evolution that a
gradualist would expect. There is no 'trend' of gradual accretion
of brain size. The range of brain sizes *within* populations of *H.
erectus*, who lived 1.8 million years ago in what is now Dmanisi,
Georgia, show the same large variations that we find in today's
modern humans. In fact Lordkipanidze et al. (2013) think
that the Dmanisi single population show the same amount of

variability as the rest of the entire early *Homo* fossil samples. They now believe that all *Homo* fossils found so far most likely represent a single, albeit diverse, species.

Excavations in Olduvai (Africa) have revealed a 1.25 million year old cranium with a cranial capacity of 1067 ml, whereas a 730,000 year old find exhibits a brain size of only 700-800 ml (Rightmire, 1990; Antón, 2004). So the brain size of a much more recent *H. erectus* was quite a bit smaller than its distant ancestor.

For Asian fossils, the story is the same as African *H. erectus*. Cranial capacity varies between 800 ml and 1250 ml. These variations, incidentally, also fall within the range and variation that we find within modern human populations, so we should not be surprised by Rightmire's findings. Ruff et al. (1997) support this picture as well. After analyzing many hominid fossils from throughout the Pleistocene, they conclude that taking into account body mass, there was a long stasis of brain mass from 1.8 million years ago to 600,000 years ago.

Based on this analysis, Rightmire (1990) warns that we should not confuse variation that is to be expected in any population of contemporaneous individuals with evolutionary change. He raises the question for paleontologists, who expect to see a gradual transformation within the *H. erectus* line, of how to explain the fact that we appear to have only one species sampled over this geographical range. He adds that as far as a trend for an increase in cranial capacity in *H. erectus*, there is little evidence for gradual, progressive change. Rightmire concedes that there is no doubt that cranial capacity has increased in hominids from the first australopithecines and *Homo* specimens, through to late Pleistocene *H. erectus*. However he is wary of graphs that draw a line between these fossil finds, as if there was a gradual and steady increase.

Although the brain size of some *H. erectus* may have been similar to those of *H. sapiens*, we should not assume that the architecture of both brains was the same. Hublin and Coqueugniot (2005) have shown that although the absolute brain size of *H. erectus* may have been almost the same at birth to that of *H. sapiens*, the *H. erectus* brain did not follow the same developmental curve. They point out that brain size at birth in primates generally equates to body size, which is the approximate case with *H. erectus* as well as in modern humans.

According to Hublin and Coqueugniot (2005), a 1.8 million year old *H. erectus* specimen aged between 0.5-1.5 years, known as the Mojokerto child, appears to have followed a brain growth pattern similar to that of a chimpanzee. Chimpanzees, according to Balzeau et al. (2005), achieve approximately 80% of adult brain size by one year old. It should be noted that the human infant brain reaches about 825 grams by the end of the first year, which is roughly the size of a fully developed *H. erectus* brain (Montagu, 1989). This indicates that the *H. erectus* brain had quite a rapid growth pattern in the first year.

Developmental trajectories have profound effects on resultant cognitive abilities. Modern children suffering from severe autism as a result of extremely rapid cerebral growth during the first postnatal year, show a range of cognitive deficits. Hublin and Coqueugniot (2005) believe that this developmental pattern may be reminiscent of the *H. erectus* growth trajectory. The excessively rapid brain growth in autistic children does not allow for normal axonal and dendritic growth, which creates, reinforces and eliminates synapses as the infant develops social and cognitive skills.

Research using MR images shows the importance of myelination of the frontal lobes during early human infancy, especially between the age of 2 months to 3 years. Most clinical cases of retardation, according to Carmody et al. (2004) are due to abnormal patterns of myelination, particularly white matter fibres linking cortical and subcortical regions of the brain. Myelination continues exponentially over the first 3 years of human infancy and has reached adult-like patterns by around 36-40 months. Carmody et al. believe that the pattern of myelination relates directly to developmental changes in human behaviour, including social and emotional aspects as well as self-recognition and the ability to engage in abstract pretend play.

Hublin and Coqueugniot (2005) surmise that it is unlikely that *H. erectus* could have exhibited mental abilities similar to those of modern humans. Using CT scans of a fossil endocast, they determined that a *H. erectus* infant had achieved 84% of adult brain size by one year old, whereas the human infant has only around 50% of adult size at this age. Naturally, this conclusion is based on the estimated expected adult brain size of this hominid, something difficult to judge due to the variation in adult body size found within this species.

Importantly though, Balzeau et al. (2005) recognize that certain traits in this young *H. erectus* have similarities with modern humans of the same age. In the earliest stages of development the cerebrum is rounded and there is a marked expansion of supero-inferior regions in the brain. Following this expansion, the *H. erectus* brain starts to flatten out, particularly in the antero-posterior development of the frontal lobes. Balzeau et al. note that the early fast growth pattern of *H. erectus* appears to have exerted significant pressure of the frontal lobes against the frontal bone. They claim that the morphology of the brain of this young *H. erectus* indicates a low development of the prefrontal cortex compared with *H. sapiens*.

In humans, the cerebrum increases in height as it grows into an adult brain and retains its rounded shape. Balzeau et al. (2005) believe that *H. sapiens* present a neotenic retention of *H. erectus* brain growth as evidenced by the shared relative development of the cerebral lobes. Humans at 10 years old have usually reached about 90% of adult brain size. This prolonged growth period in humans allows for the development of cognitive and social skills in an 'enriched environment' and Coqueugniot et al.

(2004) argue that this prolonged interaction of sensiormotor cortical areas with peripheral somatic areas is essential for the development of spoken language.

McKinney (1998) has found that mammalian brain development is restrained due to the mitotic rates of neuron production, but the modern human brain has a longer growth phase than other mammals, and this has profound cascading consequences for later brain and cognitive development. Extension of infancy and juvenile stages allows for extended dendritic growth in the cortex as well as glial cell growth and prolonged synaptogenesis. Myelination starts in the brain stem and proceeds in subcortical areas and lastly neocortical, especially prefrontal cortex, which is extended to well over 12 years of age in modern humans. This, McKinney argues, allows for more effective nerve transmission, which improves memory and allows for the greater intelligence and language skills, which emerge during human development. In particular, it seems as if the neocortical size in humans has been crucial for cognitive function, and the enlarged prefrontal area has allowed for increased short-term memory. This, in turn, has given humans greater ability for the storage and manipulation of information.

Cheney and Seyfarth (2005) remind us that the acquisition of language in infants presupposes a rich conceptual structure, which predates language and is one of its required causal engines. If the pattern of *H. erectus* brain growth was fundamentally different from human infants, it seems doubtful that this species had the cognitive apparatus to form the complex representations that underly the formation of human concepts and the acquisition of language. As argued in the next chapter, we have scarce evidence for a rich conceptual life, or an understanding of causal relations, in any of our hominid ancestors.

Neandertal Brains

Neandertals were hominids that occupied most of Europe immediately before *H. sapiens* moved into their territory and most likely caused their extinction over a very short period of time. Neandertals had brain sizes as large as, or often exceeding, those of *H. sapiens*, yet, for some, it has been a puzzle why they appeared to be far less cognitively advanced. However, after extensive analysis of 163 fossil individuals from the Pleistocene, Ruff et al. (1997) have concluded that relative to body mass, Neandertals were slightly less encephalized than modern humans. Their large brain mass is attributed to their larger than average body mass for Pleistocene *Homo*. This is the same picture that we saw with the emergence of the first *H. erectus*. Some individuals had a brain size that was almost double that of the australopithecines, but so was their body size.

Ruff et al. (1997) make the important point that Neandertals are not morphologically homogeneous, and did not evolve from a 'primitive form' into a more modern looking form. In fact, there was a long period of stasis in brain mass from 1.8 Myr up until 150 thousand years ago.

When it comes to the artifacts associated with Neandertals, the picture is once again one of conservation of form. According to Minugh-Purvis and Radovčić (2000), all of the stratigraphic levels of Neandertal sites show a "remarkable typologically consistency" of stone tool, and by extrapolation, their behavior is viewed as fundamentally the same throughout the time of these deposits. It therefore seems unlikely that their brains 'evolved' from a primitive and simple form 'into' a large brain enabling more complex behaviour.

Manzi et al. (2000) find that the European hominids, including the Neandertals, appear to have had a quite different ontogenetic pattern of growth of the brain and skull. It appears that the large brain of Neandertals caused hypostosis, defined as insufficiency of osseous development. They basically suffered from mechanical stress on the bony structures of the skull, which meant that the sutures of the skull could not complete their ossification. Manzi et al. have also found that, contrary to the European pattern, the African clade at this time shows a different pattern of development of skull and brain encephalization, further supporting the origin of modern humans in Africa.

We should nevertheless recognise that Neandertals appear to have survived quite well for many hundreds of thousands of years in Europe. As Tattersall (2006) observes, "*Homo neanderthalensis* possessed a mastery of the natural world that had been unexceeded in all of earlier human history" even though they left behind "precious few" markers of any complex cognitive behaviour. Chapter 5 examines the archaeological evidence associated with Neandertals, and the findings lead us to conclude that Neandertal did not exhibit any of the behaviours that we typically associate with humans. Just like other *H. erectus* populations, they did not ritually bury their dead[10], they were not technologically 'inventive', and they appear not to have modified the environment to maintain home bases.

4.7.3 *Homo floresiensis*

Arguing against a single line of hominids leading to *H. sapiens* are the recent and controversial finds of adult skeletons found in Liang Bua cave on the island of Flores, Indonesia. These hominids are believed to have survived right up until 18,000 years ago. Based upon a mosaic of primitive and derived features, this hominid has been given a unique species name *H. floresiensis* (Brown et al., 2004).

This species shows a combination of cranial features closest to *Homo ergaster*[11], and postcranially *Australopithecus garhi* (Argue et al., 2006). Its limb proportions are similar to those of early hominids and chimpanzees. The teeth show a mosaic of primitive and more modern features, which make it difficult

---

[10]See section 5.4 for a study of hominid burials.

[11]*H. ergaster* is the species name given to a group that are otherwise known as African *H. erectus*.

to place it clearly within any known species. *H. floresiensis* had no chin and a forehead that sloped straight back from its brow ridges, just like *H. erectus* and all of the australop-ithecines.

Due to this mosaic of anatomical traits, Argue et al. doubt that *H. floresiensis* evolved from an endemic *H. erectus* in Indonesia. Brown et al. (2004) have speculated that this new hominid may have evolved by gradually 'shrinking' from a previously full sized *H. erectus* in order to cope with the lack of resources on its island home. They also believe that the quite advanced stone tools found in the same cave as the fossil hominids were made by *H. floresiensis*, indicating a more evolved mental capacity. However, Tocheri et al. (2007) point out that primitive stone tools, similar to the first stone tools found in Africa, are directly associated with the *H. floresiensis* fossils, which suggests that the more modern stone tools found in the cave were made by *H. sapiens*.

Argue et al. (2006) doubt claims that *H. floresiensis* possessed developed cognitive abilities due to a purported prominent Brod-mann's area 10. Encompassing the prefrontal cortex and tem-poral lobes, area 10 is usually associated with higher cognitive processing, but studies of microcephalic modern humans with similar brain sizes to *H. floresiensis* (380-410 cm^3) show a great deal of variability in overall brain shape. They also find that microcephalic humans possess a highly variable cognitive and behavioural ability and they conclude that the correlation be-tween brain morphology and behaviour remains unclear.

Suggestions that *H. floresiensis* must have had the planning ability to build water craft enabling them to reach the island of Flores have also been disputed by Argue et al. (2006). While the region of Wallacia now has quite high sea levels, this region is tectonically unstable and small-scale tectonic movements in the past may have provided temporary land bridges, which would have facilitated crossings by these early hominids.

A most interesting revelation about *H. floresiensis* comes from analysis of the wrist bones of this hominid. Tocheri et al. (2007) have found that the wrist bones of *H. floresiensis* are closer in form to non-human primates and australopithecines. The derived configuration of wrist bones that we find in humans appears to have arisen by around 800,000 years ago. From this observation, we could surmise that *H. floresiensis* must have had a fundamentally different developmental pathway from humans and was most likely closer to our Great Ape ancestor.

The most important point to be made is that once again we do not find a gradual evolution of hominids as a 'trend' from the 'primitive' to the 'complex'. The evidence emerging from studies of *H. floresiensis* leads to only one conclusion: the first hominids maintained a more or less conservative phenotype for most of their evolutionary history, from 5-7 millions years ago right up to just 18,000 years ago. Most of the variation that we find in our hominid ancestors can be explained by interbreeding of highly variable morphologies of perhaps a single species of hominid or the effects of hybridization with our primate ancestor (Patterson et al., 2006; Ackermann et al., 2006). Regardless of variations in australopithecine and *Homo* anatomy, we find a picture of stasis and certainly not a case of gradual 'advance' in hominid evolution from 'primitive' to 'complex'.

4.8   Summary

I have argued that we have no reason to look for Darwinian selection pressures to explain the evolution of the hominid clade. The first ancestral hominid appears to have made a 'sudden' appearance as the result of a saltatory event that produced bipedalism. Bowers (2006) believes that the chromosomal fusion that occurred in our hominid ancestor would have easily become established in just a few generations. We saw that this chromosomal fusion was the most likely cause of the sudden appearance of bipedal hominids. We have little reason to doubt that this creature interbred with the original primate group to produce *all* of the variation that we now find in the fossil record over a five million year time span. Bowers reminds us how easily an evolutionary novelty can quickly be incorporated into the gene pool of a small single male–multiple female social structure.

According to Ackermann et al. (2006), it is generally unappreciated that hybridization often leads to novel genotypes/phenotypes as well as the origin of new species. Their studies of hybridization in extant primates reveals that hybrid skeletons are often morphologically distinct from their parent species (often being larger or smaller than expected, especially evident in the morphology of the teeth). Australopithecines show highly variable dentition, and are usually classified either as robust or gracile purely on the size of their teeth. The reduction of the hominid jaw, due to its paedomorphic origins, created dental overcrowding as is evident by finds of supernumerary teeth and molar impaction among many hominids.

It seems most likely that early hominids were one species of primate for their entire, roughly 5 million years of existence. We find little evidence of trends toward greater fitness as would be expected under a Darwinian selectionist account. Rather we find intraspecific variation that seems to have held for this 5 million years of stasis.

Selectionist accounts of evolution that assume fitness of form leading to reproductive advantage must explain how certain malfunctioning parts of the human body have remained in the population. Indeed, Lewin raises the interesting challenge for evolutionary theorists who should be able to explain

> how it was that an apelike ancestor, equipped with powerful jaws, and long, dagger like canine teeth and able to run at speed on four limbs, became transformed into a slow, bipedal animal whose natural means of defence were at best puny (Lewin, 1987).

If we need an adaptation 'story' to explain how an organism comes to have a certain function, which improves over time due to natural selection, then we also need an account of those parts that appear to have no function, or even appear to have regressed (e.g., dagger like teeth). It is interesting to note that we nevertheless have retained the neural underpinnings that drive the aggressive use of our anterior teeth. Archer (1988) has shown that the vestigial aggressive instinct in humans when we are stressed or angry can be found when we sometimes whet[12] our anterior teeth in preparation for attack. This is a purely an instinctual reaction, albeit repressed, to antagonistic encounters.

For around 5 million years, the hominid clade appears to have been simply bipedal apes. The compelling evidence from the paleontological and archaeological record reveals an astonishing stasis for this great expanse of time. The implications of this lack of any advance in the evolution of human-like cognition and language are quite profound. Importantly, we cannot find any obvious discernible advances between the cognitive abilities of the earliest *H. erectus* and those that lived up until the emergence of *H. sapiens*. The picture for the entire *H. erectus* species is one of stasis, and certainly not one of a gradual increase in brain size or cognitive ability.

It appears that the heterochronic changes that transformed the anatomy of our hominid ancestors were purely anatomical and not cognitive. However, when we come to the evolution of *H. sapiens* the picture changes. Modern humans have retained the infant bodily proportions of their immediate *Homo* ancestor into their adult life. Current archaeological and paleontological evidence points to Africa as the location where this sudden phenotypic change originated.

Although *H. sapiens* and *H. erectus* reach approximately the same body size and shape, it seems that the prolongation of fetal growth rates has led to the retention of juvenile proportions,

---

[12]To whet one's teeth is to grind and sharpen.

especially in the cranium and face. As Nishimura et al. (2006) point out, facial flattening is most likely the major factor in the ontogeny of humans that built our unique supralyngeal vocal tract paving the way for speech. Our neotenic growth patterns have meant the late closure of the sutures of the skull, which has fortuitously allowed for the retention of embryonic growth rates of the modern human brain to continue well into postnatal development.

From a behavioural aspect, we can also see that the retention of the ability to learn is an extension of the existing capability of juvenile chimpanzees. Montagu (1989) notes that the educability of the juvenile ape is much more attainable than the adult ape. The retention of juvenile brain growth rates seems to have had a highly selectable potential as we can see in the human ability to continue learning, generally throughout our lives.

In the next chapter, I present more concise evidence from archaeology, which supports the claim that modern human cognition arose only together with the crucial mutation that produced *H. sapiens*, in Africa, around 120,000 years ago.

5. Archaeological Evidence

5.1 Introduction

In this chapter, I offer archaeological evidence to support the view that what we normally think of as the 'essence' of being human arose only recently in our evolutionary history, with the emergence of anatomically modern humans. I summarize the archaeological evidence that challenges the assumptions that our hominid ancestors were making complex multi-component tools, were burying their dead, had the technological know-how to build shelters, fireplaces, or any structures that required a complex planning ability, and in general, had an advanced understanding of cause and effect.

Human beings undoubtedly exhibit the most complex social behaviour and technical ability of any animal. Tomasello et al. (2005) point out that humans not only interact with conspecifics socially, like other animals, but they also engage with others in complex collaborative activities. Humans make tools, prepare meals, and build shelters together. They also play cooperative games, enjoy dancing, and playing and listening to music together, and they often collaborate in scientific endeavours. Humans exchange love tokens and engage in rituals that celebrate various life-stages, with the most visible ritual that we recognize as human-like being burial of the dead. These collective activities are underpinned by social institutions, and also language, which helps to continue these activities across generations.

Tomasello et al. propose that only humans are biologically adapted for participating in collaborative activities where shared intentionality is involved. They note that the motivation and skills for participating in this kind of shared experience emerges in the early stages of ontogeny. Young children experience the 'we' kind of intentionality that allows them to be part of a collective human cognition. No such development occurs in any of our Great Ape cousins, neither in infancy nor in adulthood. There is little or no evidence that 'untrained' apes indulge in planning exercises, or collaborate in any meaningful way that entails a communication system (see also section 7.2.4).

Under a Darwinian adaptational scheme, we should expect a gradual increase in complexity of all of the above behaviours. Although we have little evidence for how hominids interacted socially, the little we do find does not suggest a close connection with what we recognize as 'human'. The earliest unambiguous evidence that hominids were butchering one another is seen in the disarticulated remains found in a cave in Sterkfontein, South Africa. These remains, dating to the early Pleistocene, were of australopithecine, and may represent the 'prey' of early *Homo erectus* (Pickering et al., 2000).

Even comparatively recent remains of hominids[1], which indicate dietary cannibalism, have been found in Spain, dating to 780,000 years ago. Fernández-Jalvo et al. (1999) have excavated the remains of six individuals, which had been randomly discarded along with stone tools and other nonhuman remains. Both the hominid and other animal remains show similar intensive exploitation with a preference for infant and juvenile age groups. All of the remains have signs of peeling, percussion breakage, muscle cutting, torso dismemberment and cutmarks indicating access to viscera. Evidence of fire is lacking at this site, so we can assume that all of this material was eaten raw. Fernández-Jalvo et al. suggest that, due to the similar butchering techniques, breakage patterns to extract the marrow, and identical pattern

---

[1]The earliest fossil remains of a bipedal primate date to between 6 and 7 millions years (Lahr and Foley, 2004).

of discard, these hominids viewed other hominids of their own species as food. They point out that the climate at this time in Spain was temperate and a high diversity of species is found in all levels of this stratum. It appears that this gastronomic cannibalism could not be considered part of a survival strategy due to a period of starvation.

It appears that Neandertals were not the only species indulging in cannibalism. Recent finds from the Les Rois cave site in France, dated to approximately 30,000 years, reveal an interesting picture. Rozzi et al. (2009) have excavated this site and found remains of both modern humans and Neandertals. The tools are Aurignacian and therefore attributed to modern humans (*Homo sapiens*). This site shows clear evidence that modern humans were interacting with Neandertals, but cutmarks on the bones of juvenile Neandertals indicate to Rozzi et al. that humans may have seen Neandertals as just an edible resource. They suggest that this find may throw light on how the two species were interacting, and surmise that this contact most probably contributed to the demise of the Neandertals. Analysis of mtDNA sequences of four Neandertals and five early modern humans found no evidence of mtDNA gene flow in either direction (Serre et al., 2004). However, all four Neandertals were found to be closely related. Although speculative, it would not be too presumptuous to assume that as humans moved from Africa into Europe, around 43,000 years ago (Mellars, 2006) they found a population of primates that were not recognized as 'cousins', but as another (edible?) species, with little or nothing in common as far as behavioural traits.

In the following sections, I survey the archaeological record for signs of human-like behaviour, and conclude that symbolic behaviour, including language and ritual, arose only after the emergence of anatomically modern humans around 120,000 thousand years ago.

5.2   Stone Tools

Stone tools are generally the most visible artifact in the archaeological record, and signal to many scholars the arrival of the
human line of descent. Arguing the case for a link between stone
tools and an advance in cognitive ability, Bridgeman (2005)
proposes that the making of stone tools requires the ability for
creating and storing plans for sequences of actions. This, he
believes, may be an indication that the makers of stone tools
were most likely language users. Gergely and Csibra (2005)
claim that the manufacture of stone tools required complex
skills that could not have been acquired by mere observation,
but would have required passing of this knowledge through
language skills.

In this section I survey evidence supporting the argument that
the making of simple stone tools does not necessarily signify any
major advance in cognition over and above that of our Great
Ape ancestors. Chimpanzees and orangutans are capable of
transporting materials for the making of tools for the extraction
of hidden resources, for cracking nuts, and for the making of
tools for hunting (Mercader et al., 2002; Pruetz and Bertolani,
2007; Van Schaik, 2006). We should bear in mind that stone
tools become evident in the archaeological record only after a
period of up to 5 million years of 'evolution' of our hominid
lineage. The split from a Great Ape ancestor is currently
thought by some to have happened up to 7 million years ago
(Schwartz, 1999). If we started to evolve human-like cognition
right after the split from our Great Ape ancestor, then we
should expect some change in the archaeological landscape, as
evidence of a move toward more complex cognitive behaviour.
However, the first stone tools, which are attributed to *Homo
habilis*, are found in Africa and dated to 2.6 million years old
(Semaw, 2000).

Binford and Ho (1985) point out that "the archaeological record
does not speak for itself". They challenge the old argument

that "tools maketh man" arguing that many animals use tools, but we normally do not attribute a 'culture' to them. Binford and Ho also believe that "it is clearly possible that early man was a sophisticated tool user and manufacturer and yet did not have *culture*". Byrne (2004) has excavated a Neandertal site in France (Arago Cave), dated to the Middle Pleistocene, that contains many levels of stone artifacts. She has found that all levels show a stasis of technical tradition, and that the main factor determining the type of end product was the raw material available for knapping. Byrne construes this long chronological sequence of stone tools as rational adaptation to available raw materials, rather than as strategies grounded in cultural tradition.

We find similar behaviour in extant pongid. Recent observations in a West African rainforest have discovered tool use by chimpanzees (Mercader et al., 2002). Chimpanzees in the wild currently transport raw materials from various sources in order to crack panda nuts, which provide a high calorific diet for little effort. As a result of the nut cracking exercise, there is some unintentionally flaked stone, which may offer a clue as to how early hominids first discovered the utility of sharp-edged stone for use as a cutting tool. Of course, this nut-cracking 'culture' does not imply the ability to manipulate symbols or use a language of any kind. A more radical view comes from Gargett (1993), who is inclined to dismiss claims for the arrival of "technology", when the term "breaking rocks" would suffice. He points out that the term technology is usually associated with a "fabric of meaning", and is often associated with linguistic competence. However, Gargett makes the important observation that we do not automatically associate a fabric of meaning to other animals that are able to transform their environments (e.g., beavers) using natural materials.

It has been argued that the complexity of stone tools made by ancestral hominids is not obviously advanced from those of our close Great Ape ancestors. Ambrose (2001), for example, believes that the apparent refinement of technology with the first stone tools, attributed to *H. habilis*, may be partly due to a change in the basic anatomy of the fingertips and thumbs, which led to greater flexibility and control over that available to chimpanzees. I argue that the end products of simple stone tool manufacture do not automatically necessitate an intentional mind, capable of planning, teaching, or holding a template of a desired finished artefact.

It is not known for certain who made the first stone tools. Toth and Schick (1993) point out that a robust form of australopithecine and a slightly larger brained *H. habilis* were living side-by-side around 2.5 million years ago, when the Oldowan technology emerged in Africa[2]. Most stone tools are found in areas where *Homo* remains have been discovered and these tools are around 2 million years old. Toth and Schick suggest that although these hominids had mastered some important principles of flaking stone, (like being able to recognise acute angles), we should not assume that they needed "mental templates" of the procedures involved. The cores that remained after one or more flakes had been struck off, although useful perhaps for chopping or scraping, were just the end product of the reduction process. Their shapes were determined by the size, shape, and raw material of the rock used. Nevertheless, they suggest that because these hominids were actually transporting raw materials to activity sites, they may had more foresight than that exhibited by modern apes. However, as noted above, chimpanzees have been known to transport raw materials in order to crack panda nuts. So, even to attribute foresight and planning, over and above that displayed by other Great Apes, seems unwarranted.

---

[2]The first crude stone tools, known as the Oldowan tradition, appear in the archaeological record around 2.6 million years ago and are attributed to *H. habilis*, although there are no stone tools directly associated with the fossil remains of this hominid. There has been some speculation that *Australopithecus garhi* may be a candidate for the maker of the first stone tools found in Gona, Ethiopia.

5.2.1 Chimpanzee tool use

Recent studies of chimpanzees in the wild have revealed that chimpanzees not only use tools for termite fishing and nut cracking, but also that they are quite capable of modifying and using tools for hunting. Pruetz and Bertolani (2007) have studied the hunting behaviour of a group of chimpanzees in Senegal and found that they construct tools for hunting prosimian prey. Some of these tools take up to five steps to produce, including trimming the tool to a point to be used as a spear like weapon. This type of behaviour, they argue, involves foresight and intellectual complexity, and suggests that this may have equaled the kind of capacity found in the first australopithecines. A most interesting observation is that females and immature chimpanzees appeared to indulge in this hunting behaviour even more than the adult males of the group. This suggests that we should "rethink traditional explanations for the evolution of such behavior in our own lineage" (Pruetz and Bertolani, 2007). Nevertheless, we have not witnessed chimpanzees engaging in any form of 'language' or 'teaching' when making or using tools, so it seems ambitious to speculate that our hominid ancestors were acting any differently.

5.2.2 Orangutan tool use

An interesting study of orangutans in Sumatra by Van Schaik (2006) has found that 'culture' plays a major role in the ability to learn. The orangutan is known to be normally rather solitary and socially reserved, unlike the more hyperactive and socially convivial chimpanzee. Van Schaik has observed that one particular group of orangutans in the study area appear to be intrinsically smarter than other groups living nearby. The difference between the groups appears to be in their learning culture. Only one group has learned to prepare a stick tool, which is used to extract the highly nutritious seeds of *Neesia* fruits. These seeds, which contain around 50 percent fat, are highly prized and provide most of the sustenance for the tool-using orangutans. Although orangutans normally maintain a close mother-infant bond, adults do not usually spend much time together. However, Van Schaik has found that the tool-using group have developed a more connected social network. It appears that this group has developed a social system whereby all members are exposed to learning opportunities, especially when foraging. Van Schaik surmises that this social network means that any newly invented skill would be retained and reinforced through generations.

It should nevertheless be pointed out that this group of more socially connected primates does not engage in any symbolic communication system. This fact undermines theories that posit enhanced social cohesion as the reason for the evolution of human-like communication, and other more complex cognitive traits.

5.2.3   The First Hominid Stone Tools

The oldest stone tools to date have been found in Gona, Ethiopia, and have been dated to around 2.6 Mya (Semaw, 2000). Among the finds of stone artifacts, we find evidence of stone-tool cut-marks and hammerstone fractures on animal bones. This indicates to Semaw that at this time there must have been an incorporation of substantial meat and bone marrow into the hominid diet. The 'invention' of stone tools as a butchery implement seems to have heralded an important stage in ho-minid ability to exploit new environments. Semaw points out that we find many more butchery sites in Eastern, Northern, and Southern Africa, after the discovery of stone tool flaking techniques.

Nevertheless, it is important to note that this new discovery did not equate with an 'unleashing' of novel technological skills. Most of these stone tools are produced by a 'single strike' action on a pebble, creating a cutting edge along the cleavage. The stone artefacts reveal a simplicity of form and knapping technique. Semaw believes that these hominids were mainly after sharp-edged flakes to be used as cutting implements, which is evident in the selection of finer and easy-to-flake raw material. Renfrew and Bahn (1994) consider that these early stone tools may represent just a simple, habitual act, not unlike a chimpanzee breaking off a stick in order to poke at an ant hill.

The makers of these stone-tools appeared to have quite good coordination and knapping skills, but for Semaw, these artifacts do not appear to be the result of an intended, predetermined design. He is cautious about claims that these hominids evolved their skills gradually. There are no precursors in the archaeological record of these simple stone tools, and exactly the same types of stone tools, involving the same skill levels, are found one million years later in the same regions of Africa. The stone-tools did not become more elaborate, nor was the selection of raw materials more refined during this vast period of time.

Noble and Davidson (1997) make the interesting point that once human stone knappers have mastered the technique of producing a sharp flake with a single blow, they soon recognize that an astounding variation of sizes and shapes can be made. This recognition, Noble and Davidson suggest, seems to have gone completely unnoticed or unappreciated by our hominid ancestors. The fact that early hominid tool productions were so similar enables archaeologists to place them within an industry[3].

Ambrose (2001) has shown that the assemblages of stone tools at this time reflect a least-effort strategy from the available raw materials. Most of the stone tools were derived from small pebbles of about 2.5 cm across and were too small to be flaked in

---

[3]A term used in archaeology to describe a group of tools with similar characteristics. The stone tools produced by australopithecines or *H. habilis*, are grouped together under the name Oldowan industry.

the hand. They would have been placed on an anvil and smashed with a hammerstone. I believe that this observation adds credence to the idea that the 'invention' of the first stone tool may have actually been an accidental 'discovery'. Ambrose sees little sense in assigning this technology to culturally determined stylistic traditions.

Did the first stone tools made by hominids reflect greater mental capacities than extant apes? Even after years of training, the chimpanzee Kanzi cannot produce stone flakes that resemble the quality of the first stone tools created by australopithecines. Kanzi can produce a flake when required, especially for cutting through the rope on a box containing a food reward. However, there are important anatomical restrictions placed on a chimpanzee. Ambrose points out that to cut the rope, Kanzi cannot use his immobile wrist, but has to move his whole arm, mainly from the shoulder with little downward pressure. This is the same type of movement that we see in wild chimpanzees when they are cracking nuts.

Humans have the anatomical advantage of short, straight fingers with a stout thumb and fleshy fingertips, which all aid in a precision grip. In addition, humans have a flexible wrist, which gives greater flexibility during tool use. Some australopithecines and *H. habilis* appear to have had more human like fingertips and thumbs, so Ambrose thinks that this may account for the slightly more refined stone tools over and above those that are able to be produced by chimpanzees.

The sudden appearance of thousands of well-flaked stone artifacts 2.6 million years ago subsequently supported the rapid spread of hominids to other regions. From then on, a technological stasis ensued. This pattern of the novel emergence of technology, and subsequent stasis of form, does not support a theory of gradual change due to natural selection of cognitive improvement over time. Rather, it supports a theory of sudden appearance of novelty, which spreads rapidly through populations, and is then stabilized by its inclusion in a repertoire of enhancements of survival strategies.

5.2.4 *Homo erectus* Stone Tools

H. erectus[4] in Africa is now accepted by most scholars to be
the immediate ancestor of *H. sapiens*. Although these hominids
emerged around 1.8 million years ago, they continued to make
the very simple stone tools described above for 300,000 years.
Around 1.5 million years ago, a new type of stone tool, known
as the Acheulean hand-axe[5], appears among the assemblage in
Africa. This stone tool, in its most symmetric production, has
a distinctive teardrop shape. The hand-axe was produced by
H. erectus and remained unchanged in form from its emergence
1.5 million years ago right up until 300,000 years ago (Ranov
et al., 1995).

The earliest evidence for hominid occupation in Central Asia
has been found at Kuldara in Tajikstan (Ranov et al., 1995).
Here the stone tools are dated to 850,000 years ago and are
attributed to *H. erectus*. More artifacts have been found in later
deposits dated to around 130,000 years ago. This lithic industry
is described by Ranov et al. as pebble tools, which have neither
shaped cores nor bifaces. This site offers no evidence of spatial
organization or the use of fire. Interestingly, the supposed more
'advanced' Acheulian lithic industry, which had its origin in
Africa around 1.5 million years ago, seemed not to have reached
Central Asia.

In China, the stone tools associated with *H. erectus*, and dated
to around 1 million years ago, belong to the lithic industry
that first arose in Africa 2.5 million years ago (Ranov et al.,
1995). Throughout China, up until about 160,000 years ago, the
lithic industry associated with *H. erectus* remained unchanged
or non-existent. Within the *H. erectus* clade, technological
variability seemed to decrease outside its area of origin, Africa.

---

[4]There is ongoing controversy over which species of *Homo* is the immediate ancestor of *H. sapiens*, (e.g., *H. ergaster, H. heidelbergensis, H.
antecessor*).

[5]The classic Acheulian hand-axe was flaked on both surfaces of a flaked
stone producing its symmetrical shape (Renfrew and Bahn, 1994).

Unlike modern humans, our ancestor species seems to not have had the cognitive capacity to adapt technology to different environments.

It seems that tool technology needs constant and continual cultural transmission in order to survive as a useful trait. Toth and Schick (1993) suggest that the technology may have died out after one generation as *H. erectus* populations traversed landscapes lacking large raw materials suitable for making hand-axes. They argue that the cognitive processes, especially linguistic capability, of these hominids were not enough to maintain any technological traditions. At around 500,000 years ago, it appears that a "new wave" of *H. erectus* emerged from Africa and spread into Europe, once again bringing the Acheulian hand-axe with them.

It is often suggested that the arrival of stone tools, linked to hominid remains, may signify a first step into higher cognitive behaviour involving planning or symbolic representation. For example,

> the sophisticated teleofunctional understanding of tools during hominid evolution led to complex skills of tool manufacturing that became practically impossible to acquire based on observable evidence through existing forms of social learning (Gergely and Csibra, 2005).

Bridgeman (2005) also maintains that a planning capability for creating, storing and executing plans for sequences of actions, which has evolved in primates, was exploited by hominids to apply to communicatory acts. He believes that the ability to generate a sequence of actions, which are organized in a hierarchical fashion, with smaller tasks embedded in larger ones, grew in complexity and eventually led to producing a sequence of words. Gergely and Csibra say that the sophistication of the hand-axe reflects a preconceived design, which in turn reflects a new stage of hominid cognitive abilities. Savage-Rumbaugh

(1994) also asserts that there is a link between language and the making of stone tools, which she thinks involves the ability to assemble something in a hierarchical manner. Several scholars have challenged these assumptions. For example, Ambrose (2001) has shown that Acheulean hand-axes are more likely to be the "unintended byproduct of nonstylistic factors rather than intended target types". When raw materials are particularly scarce in a region, a cutting tool will need to be re-sharpened rather than discarded. In areas of scarce resources, he observes how a simple cleaver with little reworking becomes a teardrop-shaped hand-axe after several bouts of resharpening.

Jones (1979) has found that the properties of the raw material available for knapping are the most important factors determining the technological complexity of a stone tool. The mechanical properties of the raw materials ultimately determine the ease of flaking, the final shape, and the apparent refinement of the cutting or slicing tool. Some raw material is brittle and produces an initial sharp edge, which is very sharp, but is easily blunted. This sort of material is easily resharpened and may look as if it has been produced using a more complex technology of rework. Other raw material, like basalt, is efficient for cutting and skinning and lasts a long time, but is not easily resharpened. Therefore, basalt tools may appear to lack a refinement of technology.

There is no progressive trend in the manufacture of these tools. The apparent complexity of the finished product is simply determined by the level of difficulty in achieving a sharp edge with the raw material at hand. Moreover, studies of the brain using positron emission tomography (PET) reveal that stone tool making should be characterized as a complex sensorimotor task involving spatial-cognitive skills, rather than complex cognitive skills involved in higher-order thinking and planning (Stout et al., 2000). Based on this evidence, it seems premature to link the shape, size or apparent refinement of a stone tool to any trend for progressive complexity in hominid cognition.

Ambrose (2001) concludes that both the cultural and the cognitive capacities of these hominids may have been substantially overestimated. These findings are backed up by Soressi (2004), who believes that changes in tool-making were not the result of a tendency toward complexity, but a result of the need to exhibit more opportunistic behaviour as competitive pressures prevailed [6].

Gergely and Csibra (2005) suggest that a system of communication was needed to pass on relevant tool-making information to the next generation. This assumption raises a further serious problem. As outlined above, the archaeological evidence suggests that abilities for tool making did not survive migration, and thus this purported ability to communicate a "teleofunctional understanding" of the manufacture of stone tools was not in fact part of pre-human hominid culture. Migrating populations need to have a system whereby one generation passes on knowledge to the next, especially when the materials that support that culture are not readily available.

5.2.5 Neandertal Stone Tools

The hand-axe first attributed to *H. erectus* in Africa was widely distributed across Europe, and its more refined and symmetrical forms are usually attributed to the work of Neandertals, who some believe show an advance of cognitive ability above that of other *H. erectus*. For example, d'Errico et al. (2003) believe that due to the purported refinement of these hand-axes, *late* Neandertals were much more culturally advanced than usually assumed. Moreover, they claim that Neandertals most likely had some form of language and that they may have contributed to the languages of *H. sapiens*. They base their assumptions on the time of arrival of modern humans in Europe at 37,000 years ago. Artifacts including personal ornaments found before this time, they believe, were produced by Neandertals and this, they argue, indicates that Neandertals were capable of

---

[6]See section 5.7 for further discussion on this point.

symbolic thinking. However, recent advances in radiocarbon dating methods put modern humans in Europe much earlier than previously believed, so it seems that artifacts attributed to the Neandertals were more than likely the productions of *H. sapiens*.

Mellars (2006) points out that previous radiocarbon dates were calculated on the assumption that the original proportion of the key isotopes of carbon in the Earth's atmosphere, used for dating material, have remained constant over the past 50,000 years. This assumption has recently been overturned. Radiocarbon samples from deep-sea sediments and ice-core records reveal a sharp increase in cosmic radiation during the time of modern human dispersal throughout Europe. This radiation increased the carbon-14 content of the atmosphere and accordingly changes the calibration curves. This in turn means that a measured radiocarbon date of around 40,000 years ago translates into an actual date of 43,000 years (Mellars, 2006). This new information puts modern humans not only into western Eurasia earlier than previously believed, but also shows that they overlapped with Neandertals for much less time. In western France, Mellars believes the overlap may have been as little as 1000 years. He has little doubt that this new culturally advanced and biologically modern population from Africa rapidly replaced the European Neandertals. Artifacts that were previously attributed to Neandertals can now be shown to belong to these earliest modern humans.

Johnson (1989) argues that despite their large brains, Neandertals did not organize the world in the same way as modern *H. sapiens*. Their stone tools were not organized into "tool-kits" but rather as "disposal kits". Along with their pattern of food consumption, their tools were for immediate use rather than exhibiting future-oriented behaviour. According to Johnson, this reflects a mental capacity too limited for long term planning or organization, and most likely not requiring the capacity for symbolic communication.

5.3   Language and Stone Tools

The link between tool-making and linguistic competence presents
a problem for Gibson and Ingold (1993), who point out that
we really have no idea of what tool-making demanded in terms
of linguistic competence. Moreover, recent experiments using
PET to map the areas of the brain that are active during the
making of simple stone tools reveal no evidence for a high-order
cognitive component in this task for an experienced knapper
(Stout et al., 2000). Rather, there was a high activity in regions of the brain associated with motor and somatosensory
processing, especially those areas involved in complex spatial
cognition like vision, touch and sense of body position and
motion. Their study failed to provide any evidence of a distinct,
shared neuroanatomical substrate for stone-tool manufacture
and language.

One could perhaps at this point also raise the question why
only one member of the primate order in the animal world is
purported to have evolved language as a result of the ability
to perform seemingly complex action plans. Why not beavers
and birds or even spiders? Parker and Milbrath (1993) argue
that the capacity for higher intelligence, as well as language,
evolved as adaptations for planning. Planning requires procedural knowledge, which is employed in problem-solving. Knowing
how to make a stone tool requires procedural knowledge, which
involves representation of action schemes. This procedural
knowledge may be acquired more or less automatically due to
trial and error practicing, and many skills thus embedded may
be performed unconsciously and are not communicable. These
skilled motor sequences, according to Parker and Milbrath,
do not require a symbolic representational system beyond the
awareness of a goal. From about 2 to 5 years old, human children begin what Parker and Milbrath refer to as declarative
planning, which is associated with abstract symbolic behaviour
involving drawing pictures, playing make-believe games, and
forming sentences. This transition from automated procedures

to intentional expression underpins the development of language. Our hominid ancestors appear to have reached only the procedural level of planning, as evidenced by the archaeological record. Moreover, the trajectory of *H. erectus* brain growth followed a pattern similar to chimpanzees, which is more or less complete by one year old (Hublin and Coqueugniot, 2005). This rapid brain growth in the first year of infancy would not have allow for the axonal and dendritic growth, which happens after the first year in human infants. In human infants, this altricial brain growth reinforces and eliminates synapses, which is linked to the development of social and cognitive skills.

Bickerton raises doubts about any claims for a gradual evolution of language over time, as they are at odds with the archaeological record. For example, the tear-drop shaped Acheulian hand-axe, known for its distinctive shape, and associated with *H. erectus*, remained unchanged for one million years. It is found wherever *H. erectus* roamed, which in fact covered a distance of 10,000 miles from Africa to Japan. He asks

> is it possible to think of any *sapiens* innovation that has traveled for the best part of 10 thousand miles without undergoing the slightest change?.

and that

> the mismatch between the fossil and archaeological records forms an acute embarrassment for those who believe that human cognitive capacities, including language, developed gradually (Bickerton, 2002).

Because of its symmetrical appearance, the hand-axe is often advanced as evidence for a mental image of a desired form that exists as a template the stone knapper uses to guide his actions toward the finished product. Davidson and Noble (1993) have challenged this line of thought with an interesting modern day experiment. Francois Bordes, an archaeologist skilled at tool making, and Irari Hipuya, from the Highlands of Papua New Guinea, unversed in the position that hand-axes play in archaeological controversies, both set about making stone tools. By the end of the knapping session, Bordes had a hand-axe and discarded flakes, and Hipuya had two piles of stone tools, the flakes and the remaining cores. The resultant forms of "useful" tools were different in the minds of the two knappers. Davidson and Noble use this example to demonstrate the "finished artefact fallacy", which is the belief that the final form of a flaked stone artefact, as interpreted by archaeologists, was the intended shape of a "tool". The "links" that archaeologists see in the production of stone tools are those of present-day analysis, and not those contained in the minds of their hominid makers. They have similar reservations for what many archaeologists believe indicate a more advanced technology in the Levallois[7] technique and the Mousterian[8] assemblages associated with the Neandertals. The variety of forms in these assemblages are attributed to the availability of stone raw materials in a particular area of foraging. The variability is most likely related to production contingencies rather than "planning". Where raw materials are not abundant in a particular region, retouch of tools in a mobile foraging group would be necessary.

Chase (1990) supports this view arguing that the Mousterian lithics attributed to Neandertals were not the result of deliberate fashioning of stone into tools, but rather the result of repeated retouching of flakes to sharpen their edges until the point where the piece of stone became too small to grasp efficiently. This finding, as Chase points out, has significant implications for symbolic behaviour and cognitive and linguistic abilities in Neandertals. Moreover, experiments have shown that elaborate stone tools can be produced without the use of language. Tattersall (2001) reports on an experiment with un-

---

[7]The Levallois technique is thought to involve the preparation of a core in such a way that large flakes of a predetermined size and shape could be struck off (Renfrew and Bahn, 1994).

[8]The Mousterian assemblages have been associated with Neandertals (c. 100,000–40,000 years ago) and have been described as "tool-kits" for different functions by some or just simple expedient simple tools by others (Renfrew and Bahn, 1994).

dergraduates who were divided into two groups, one half taught to make a stone tool by verbal explanation, the other by silent example. The two groups showed no difference in efficiency or speed of acquisition of the toolmaking skills. Tattersall's experiments are corroborated by Bisson (2001), whose novice flint-knappers were shown to consistently adhere to two rules. Blunt working edges were maintained to facilitate prehension of the stone flake, and the longest edge was chosen for accessibility to retouch successfully. The resulting working tool was not the result of 'imposed form' but the emergence of a structure necessitated for ease of manufacture.

Wynn (1993) has analysed the type of sequence building recognizable in sport and infant play and communication, as well as adult tool-use and tool-making. He is pessimistic about the relationship between tool making and language, and he believes that the potential of prehistoric tools to inform us about language is weak. Rather, he argues that tool-making constructs its own sequences through spatial and temporal continuity. All animals need motor memory to store these sequences, which often becomes automated, but we cannot show that these motor sequences bear anymore than a superficial resemblance to syntax. In addition, the hierarchical nature of language is not evident in the sequential strings of motor actions involved in tool-making. Wynn suggests that long-term memory capacity allowing for the sequencing of steps involved in tool-making may have evolved before language. He concludes that "the tools themselves cannot answer this question for us. They can inform us about a few general cognitive abilities only; the evidence for language must come from elsewhere".

5.3.1   Anatomically Modern Humans' Tools

The Upper Pleistocene, the era of anatomically modern *H. sapiens*, reveals a significant change in the archaeological record. Davidson and Noble (1993) convincingly argue that new artefacts were the product of deliberate planning for a desired result rather than as a product of the mechanical constraints of the technique of manufacture. Significant novelties have been found in southern Africa and the Sahara region in northern Africa, coinciding with the arrival of modern humans. Many stone artefacts have predetermined geometric shapes that can only be the result of ability to make tools to a desired shape. In addition, there appear many objects that were shaped by cutting and grinding rather than flaking, further indicating a planned sequence to achieve a desired form. Many of these objects were formed from new materials like bone, ivory or antler, materials not used by any of our pre-human hominid ancestors.

Recent discoveries in southern Africa (Pinnacle Point) reveal that early modern humans had the technological capability for not only making fire, but for using fire to heat-treat stone in order to make it more workable. Brown et al. (2009) have discovered finely worked tools made from silcrete, a material not normally easily worked into a stone tool. However, when heated, silcrete turns from its usual grey colour to a red colour with a vitreous luster. The stone recrystallizes during controlled heat treatment and becomes much easier to flake. The tools date to at least 72ka, illustrating to Brown et al. that early modern humans had reached a sophisticated knowledge of fire and an elevated cognitive ability. Other archaeological material found in southern Africa at this time also points to a sophistication of behaviour not found in any other hominid species (Jacobs et al., 2006). This advanced knowledge of fire making, and the ability to modify poor material in order to make sophisticated stone tools, was more than likely instrumental in the success of modern humans, as they moved out of Africa into colder northern European climatic conditions. This migration most probably led to the demise of the Neandertals, who exhibited little ability for this level of planning.

Davidson and Noble (1993) take the position that it is language which made this level of planning possible, entailing the ability to reflect on one's desires and produce artefacts of that reflection. They suggest that the physiological capacity for articulatory control in our ancestral hominids was recruited to express the discovery of the symbolic potential of meaningful signs. Not only do we see the emergence of symbolic thought, but the recognition of principles behind these new designs. Toth and Schick (1993) believe that this new threshold in the process of human invention shows a move away from mere observation for learning, to the understanding of mechanical devices and the rational manipulation of them. These novelties, like the bow and arrow and the spear thrower, "bespeak a burst of new, diverse ideas and experimentation with new materials and novel techniques".

5.3.2 *Homo sapiens* Artefacts and the Emergence of Symbolic Behaviour

The earliest fossil *Homo sapiens sensu stricto* have been found in Africa and the adjacent region in the Levant (McBrearty and Brooks, 2000). Anatomical evidence for the emergence *H. sapiens* in Africa, is supported by genetics, which places the ancestor of all modern humans in Africa (Pearson, 2004). The archaeological record tells us that most of what we recognize as fully modern human intentional and symbolic behaviour, arose in Africa and nearby western Asia, between 100,000 and 135,000 years ago. The emergence of modern human-like behaviour in Africa was way before the so-called 'human revolution' in Europe, which is evident from around 50 thousand years ago. McBrearty and Brooks interpret the evidence as a 'speciation event', which produced modern humans from an archaic form of hominid in Africa. The capacity for modern human behaviour is evidenced by the ability for planning, the use of sophisticated technology, and the extension of resource use. New behaviours are evident in disparate geographical and climatic regions in Africa. From its beginnings, cultural and technological change seems to have dispersed in a step-wise fashion, as it spread gradually throughout Africa, and then into Eurasia. This step-wise pattern of cultural and technological change is something that we recognize throughout the history of only one species of hominid, *H. sapiens.*

Musical Instruments

Taking a gradualist approach, d'Errico et al. (2003) argue for a remote chronology for the non-instrumental exploitation of sound and movement, which they believe may have enabled speech and song. They also suggest that proto-musical aptitude may underlie the evolution of conscious thought. This argument is formed from the belief that the sophistication of musical instruments in the form of pipes that are found in Europe around 35,000-30,000 years ago must be far removed from the origins of musical ability. They naturally build their arguments under the assumption that all human abilities evolved in a gradual trend toward complexity. The "cultural explosion" model sits uncomfortably with d'Errico et al., as this approach forces them to compress major conceptual advances into too short a period of time.

Many other archaeologists have attributed finds of punctured bone to hominids, but d'Errico et al. have soundly discounted these finds as merely carnivore activity. Rather, they argue that musical pipes that are obviously manufactured by modern humans must have had simpler, precursor forms. The notion that pipes 'built' by carnivores and subsequently utilized as musical instruments by modern humans seems to me to be a more feasible explanation, and it would avoid the problem of needing pre-cursors to these musical instruments, which needed to be 'invented' by ancestral hominids. D'Errico et al. admit that the Middle Paleolithic has so far not yielded any evidence of sophisticated artifacts, but nevertheless hold that this seemingly sophisticated technology must have arisen gradually.

Clothing

All of the Great Apes seem to manage to cope with various climatic conditions without the need for coverings. There is no direct archaeological evidence that helps us to point to when hominids started to wear clothing. However, indirect evidence, in the form of the DNA sequencing of body lice, may help with this question. Kittler et al. (2003) have sequenced the DNA of body lice that feed on humans and have found that this species evolved only at around the time of the emergence of *H. sapiens* in Africa. Body lice only live in clothing although they feed on the hairless parts of the body, whereas head lice live mainly in the scalp hair and feed on the scalp. It is quite possible that the neotenous emergence of *H. sapiens* heralded the loss of exuberant body hair that we find with other primates. Modern humans, with their new-found cognitive abilities, most likely 'invented' clothing for protection against the elements. This innovation may also have been helpful when modern humans moved out of Africa and into the much colder climates of Europe.

Ornamentation, Ritual Burial and Tools

Skhul, a site in western Asia (now Israel) and the North African site of Qued Djebbana (Algeria) are quite distant from the seashore, but nevertheless have delivered finds of perforated marine shells that indicate use as beads by modern humans (Vanhaeren et al., 2006). The assemblages indicate deliberate selection and transport to these regions. They can also be interpreted as the ability for symbolic behaviour among these peoples. Blombos Cave in South Africa has been within one kilometer or less of the sea-shore of the Indian Ocean at 50ka, 80ka and 100ka during the Middle Stone Age. Dating of a tooth has shown that modern humans were using this cave around 98ka. Levels dated from 84ka to 72ka reveal evidence of large hearths and innovative behaviour expressed in bifacially worked

foliate points, bone tools, engraved ochre, and shell beads
(Jacobs et al., 2006). Important finds in the Blombos Cave are
engraved bone fragments together with engraved ochre, which
d'Errico et al. (2003) interpreted as the most irrefutable and
oldest evidence of symbolic behaviour in humans. They point
out that no functional interpretation of these engraved lines can
be deciphered, leading them to conclude that these patterns
must have represented some sort of abstract thought. Many sites
have an abundance of red ochre showing signs of scraped surfaces
indicating their use as coloring pigments. Other large pieces of
ochre have been deliberately incised with repetitive geometrical
patterns from which we can infer some sort of symbolic or
ceremonial activity (Mellars, 2005). Large amounts of red
ochre are also associated with ceremonial burial. Henshilwood
et al. (2011) have dated ochre-rich mixtures contained within
abalone shells to 100,000 years ago. The mixture consisted
of ochre, bone and charcoal, and together with the use of
grindstones and hammerstones, indicate that humans living
at this time had a sophisticated understanding of materials
and their transformation using stone tools. In many southern
Africa sites, we find an abundance of carefully perforated shells,
which again had been imported from long distances (Jacobs
et al., 2006). Their use as personal ornaments is confirmed
by unambiguous indications of elaborate ceremonial burials
associated with a range of these shells. By 50ka, new forms of
skin working technology appear along with highly specialized
geometric blades, used as insets in multi-component hafted[9]
tools, together with intricately shaped barbed bone points. High
quality raw materials had been deliberately transported from a
distance of at least 20 km. Recent discoveries of fire-hardened
tools in southern Africa, dated to at least 70ka, illustrate an
advanced understanding of the properties of fire for changing the
nature of raw materials in order to make them more workable
(Brown et al., 2009).

We cannot say whether these first modern humans had language,

---

[9]Hafting involves fitting a handle to a cutting edge.

but we can be sure that they were engaging in symbolic behaviour, making it more than likely that some form of language was used.

5.4 Burials

We usually associate ritual burial of the dead with the concept of respect for the deceased and some sort of emotional response to that death. If grave goods are also found in the burial site, then we may infer that an afterlife was thought possible for the population of that time. Gargett (1999) points out that burial practices of our hominid ancestors are a crucial factor in the debate over the origin of modern humans. The earliest burial of *Homo sapiens* in Africa is that of an infant who was buried 105,000 years ago, with a single *Conus* shell, which is interpreted as a pendant (McBrearty and Brooks, 2000).

However, it appears that burials of our hominid ancestors are few and far between. In fact, Gargett (1989) has questioned many inferences regarding the authenticity of hominid mortuary practices, which he believes have been "incorporated into scholarly and popular view of Neandertal without, it seems, any serious criticism". He points out that we have an abundance of burials linked to the emergence of *H. sapiens*, which we recognize as evidence for humanlike consciousness or spirituality. We are reminded though, of the readily observable cultural disconformity at the boundary between Neandertals and *H. sapiens*. Most claims for purposeful burials of our hominid ancestors have been disputed by Gargett as they can be equally well explained as burial by natural processes. This observation is supported by Walker and Shipman (1996), who note that around 100,000 years ago, we find unequivocal evidence for burials that we recognise as truly human. These burials contain only the remains of anatomically modern humans.

Due to the lack of evidence for burial before the emergence of *H. sapiens*, it is entirely possible that our hominid ancestors retained the same attitude to the death of kin as that seen today with other primates. Observations by Archer (1998) reveal that baboons may carry around a dead and rotting carcass of their infant for several days, leaving it behind while they forage, before finally abandoning it. Chimpanzees appear to carry their dead infants around for only a day or less before abandonment. Archer suggests that other primates, apart from humans, do not have the conceptual level to understand the nature of death.

5.4.1   Neandertal Burials

Early archaeologists have proposed several criteria from which we could infer that burial had taken place. We should expect to find a dug grave with some sort of protection of the corpse. The position of the body is normally flexed as in sleep, and we usually find some sort of grave offerings indicating symbolic behaviour. Gargett (1989, 2000) believes that all of the so-called 'burial' sites of Neandertals were (and still are) described to 'fit' a certain picture.

McBrearty and Brooks (2000) believe that this view arose from a "eurocentric" archaeology, where in the past, workers were building their theories under the assumption that modern behaviour arose in Europe 40-50ka. In addition, Noble and Davidson (1997) claim that many of the early excavations of Neandertal skeletons were carried out before archaeological methods and the subtleties of interpretation were refined.

Many proposals for Neandertal 'burial' sites have been severely dealt with by Gargett (1989, 2000). For example, the La Ferrassie site contains nine mounds, which were described as metre-sized 'cones'. The remains of an infant Neandertal were found under one of these mounds, the other mounds being empty. These types of mounds are found elsewhere (e.g., British Columbia) and are known to be a permafrost feature. They were not created by "some mysterious mortuary ritual" (Gargett, 1989).

The Old Man of La Chapelle-Aux-Saints

One particular and oft cited 'evidence' of a 'caring' species of hominid is the purported burial of the "old man" of La Chapelle-Aux-Saints. This Neandertal 'burial' site was excavated in 1908 and yielded a nearly complete skeleton of an old man in a flexed position. Gargett (1989) recounts how it was assumed that the burial position of the old man, together with what was considered a purpose-build pit, was clear evidence for ritual care of the dead by Neandertals. Some bovid bones and reindeer vertebrae, as well as two stone tools, were found among the sediments in which the skeleton was buried. These were considered by the excavators to be grave offerings and evidence of magical ritual. However, Gargett is able to 'redescribe' this purported 'burial' site to claim that Neandertals did *not* bury their dead in a ritual manner. The old man was certainly found in a flexed position, as in sleep, but this could mean that he died in his sleep, perhaps of cold. The items that were alleged to be grave offerings are not found in the same stratum as the skeleton, most being in the level above the head of the skeleton. Gargett believes that any inference for grave inclusions is questionable. It also appears that the depression in which the body was found was not filled by sediment in a single event, which we should expect if the body had been covered by ritual burial. The depression lies at a point where the slope of the stratum changes from steep to level then steep again. There is no evidence to suggest that this depression is other than a natural feature and not a purposefully dug burial pit.

Ritual burial with animal bones

Located in Uzbekistan, the Teshik-Tash site contains a partial skeleton of a 12-year-old Neandertal together with goat horns that were described as having been placed point down in a circle. Once again, Gargett (1989) dismisses the claim for ritual burial based on the fact that all evidence points to the 'ritual' assemblage being the result of predator activity. The faunal assemblage at Teshik-Tash comprises 85% goat remains, and goat horn is the part of a goat skeleton most often preserved. Claims for a Neandertal "cult of the cave bear" have also been questioned as excavations reveal that a 'natural' accumulation of the preserved bear crania is the more likely explanation. Dense bones have a higher probability of surviving in caves and horns have the best survival rates of all.

Burial with Flowers?

Perhaps the most often cited example of Neandertal sensibility is the Shanidar cave in Kurdistan, where claims are made for ritual burial with flower offerings . Noble and Davidson (1997) point out that the word 'burial' is still used to describe this site despite the fact that it is generally accepted that these Neandertals died as a result of the cave roof collapsing on them. Gargett (1989) reports that initially, this 'burial' site was deemed to the result of burial by rockfall, but several years later, after the discovery of flower pollen in the vicinity of the 'burials', it was claimed that an individual had been buried with flowers.

However, there are many ways by which pollen can be transported into caves. Gargett proposes that this particular pollen, or even the flowers, could easily have blown on the wind into the cave. There are also numerous fossil burrows around the skeletal remains. Most rodents build their nests out of vegetation including flowers, so that this pollen could easily have been carried into the cave by nest-building animals. Noble and Davidson also point out that these Neandertals may have actually transported the plant material into the caves for bedding, a practice akin to nest-building. Insects and birds are also common carriers of pollen on their feet and in their fur and feathers.

5.4.2 Final Words on Hominid Burials

All of the above examples for Neandertal burial are, as Gargett (1989) puts it, "elaborate arguments from tenuous premises". His re-appraisal of these 'burial' sites demonstrates that, even in the case of recent discoveries, there is little or no evidence for purposeful burial or any ritual behaviour associated with death. He agrees with the observation that "workers' expectations weigh heavily on the way they 'see' the fossil record". His propensity is toward a discontinuity of behaviour, as well as morphology, with the origin of *H. sapiens*. Any shared behavioural characteristics, he believes, merely attest to a similar point of origin of both species, which we now know was at least 500,000 years ago[10].

Gargett also considers the scarcity of intact Neandertal skeletal remains indicative of the unlikelihood of purposeful burial. He reminds us that complete Neandertal skeletons are found only in very few places. If Neandertals were burying their dead, surely there should be a greater abundance of burial sites, perhaps in the same numbers as their contemporaries, *H. sapiens* (Gargett, 1989, 2000).

McBrearty and Brooks (2000) point out the significant lack of grave goods where we find the buried remains of Neandertals. They suggest that the reason we find so many buried Neandertals fossils may simply be due to the hazards of living in caves. They also point out that Neandertals may have just buried their dead in caves for hygienic reasons. It is well known that in the Levant, where we find some of the earliest burials of *Homo sapiens* dating to 90-120ka, residence in cave sites alternated at different times between modern humans and Neandertals. Modern humans associated with Ethiopian fauna ventured into the Levant around 100ka, but appear to have retreated back

---

[10]The most recent extraction of mtDNA from several Neandertal fossils unequivocally separates human and Neandertal lineages, which diverged as long as 660,000 years ago (Green et al., 2008).

into Africa due to climatic change. At this time Neandertals expanded into this region. McBrearty and Brooks suggest that a certain amount of acculturation may have taken place, and find it quite plausible that Neandertals learned the practice of burying their dead from modern humans.

The picture of our hominid ancestors as far as disposal of the dead is quite different from that of modern humans. Burial of the dead is a derived characteristic manifested only by anatomically modern *H. sapiens*. Chimpanzees do not appear to grieve or even have an awareness of death (Archer, 1998; Falk, 1992), and we see little, if any, evidence of ritual burial of our hominid ancestors. We therefore have little reason to think that they experienced the emotions in modern humans that normally accompany the death of loved-ones.

5.5   Fire

Wuethrich (1998) asks the question "when and where did our human ancestors stop running from fire and start guarding and preserving it as a vital tool for survival?". The use of fire for warmth, cooking and as protection against predators seems to be an obvious advantage that humans have gained over other animals. As Bellomo (1994) points out, the adaptive advantages given by the controlled use of fire has many evolutionary implications for hominids. Fire can be used to greatly refine the effectiveness of stone tools and also to preserve food. We have to ask the question of when did hominids make and control fire? As with hominid burial sites, many of the so-called fireplaces that hominids were supposed to have built, can be explained by alternate scenarios. Most are due to natural causes. The picture of our hominid ancestors sitting around the campfire, perhaps enjoying a meal and recounting the day's hunting as well as planning tomorrow's activities, which naturally assumes the use of language (e.g. Bridgeman, 2005), is very difficult to pin down in reality.

African Hominids and Fire

Finds in some sites in Africa, that were previously thought to be hearths, turn out to be the result of the burning of stumps in bushfires (James, 1989). Fired clay areas surrounding burned stumps disintegrate due to erosion and these clasts have been interpreted as hearths, presumably made by early hominids. Often, these clay deposits are the result of termite activity, which when burned in a bushfire, give the appearance of a man-made hearth. Several quite thick cave deposits, which were previously assumed to be ash layers created by early hominids, were, upon further examination using chemical tests, shown to be the result of burned bat guano. Apparently, lightning strikes that are attracted to updrafts from caves are a common occurrence today and are thought responsible for igniting the bat guano. Bellomo (1994) concludes from his studies of early Pleistocene archaeological sites in Kenya that hominids were not using fire for anything other than perhaps a source of protection from predators or a source of heat and light. He finds no evidence of cooking or thermal alteration of flaked tools. At the time of the emergence of *Homo sapiens* in Africa, however, an abundance of hearths, ash and charcoal is evident (James, 1989).

East Asian *Homo erectus* and fire

Early twentieth century excavations of the Zhoukoudian site in China (dated to 500,000 years ago) produced evidence for hominid occupation. This included *H. erectus* fossil remains, stone tools, and the remains of animals (including some burned bones), thought to reveal the diet of these hominids. Binford and Ho (1985) suggest that most archaeologists at this time automatically assumed that finds of this type were the result of the behaviour of 'man the hunter'. They were the first archaeologists to challenge early interpretations of this site. Archaeologists had decided that the most plausible explanation

for this 'advanced' lifestyle was enabled by symbolically motivated behaviour. Reanalysis of this assemblage by Binford
and Ho has revealed that the early workers at this site were
overly ambitious with their conclusions, and the evidence for
the early use and control of fire is tenuous. The burned bones
that were thought to have provided the key evidence for the
hunting and cooking of prey by early hominids, are now shown
to actually be the result of a discoloration of mineral origin.
Even if burning did take place, this happened long after the
bones had dried out, suggesting that bushfires had been responsible, rather than deliberate burning in a cooking process. In
addition, the six metre thick deposits that were assumed to
be compressed ash contain absolutely no evidence of hearths.
Moreover, the bone deposits, including those of hominids, are
consistent with accumulations formed by hyena activity.

It appears that *H. erectus* was not a hunter, did not build campfires for cooking, warmth and protection from predators, but
was more like an opportunistic scavenger. Further analysis by
Weiner et al. (1998) has confirmed the findings of Binford and
Ho. The putative hearths contain no charcoal, and although
there is some association of burned and unburned bones with
stone artifacts, they appear to have been deposited secondarily
by water. These deposits are found in a location where there is
no evidence of the insoluble fraction of wood ash.

European Neandertals and Fire

It appears that Neandertals were using fire in Europe as judged
by the number of burned bones, but only in the upper Pleistocene do we find structured hearths. James (1989) concludes
that behavioural complexity involving the controlled and regular
use of fire in a systematic way, indicating hunting and the use of
home bases, is more than likely unique to *H. sapiens*. There is a
possibility that the Neandertal populations in Europe may have
even acquired some of their seemingly advanced behavioural

capacities as the result of contact with early modern *H. sapiens*.
Chimpanzees can be taught by humans how to control and use
fire, so it's not out of the question that Neandertals could have
acquired this skill. Valladas et al. (1988) have found that early
modern humans were occupying caves in Israel around 90ka.
However, Neandertals, who may have migrated east out of Europe when glacial conditions covered much of their occupation
areas, had occupied the same caves around 60ka, some thirty
thousand years *after* early modern humans. Valladas et al.
think that a phylogenic relationship between these two groups
is untenable. We cannot however discount altogether the idea
that a certain amount of acculturation could have occurred,
especially in tool-making, the building of hearths, and possibly
the burial of the dead for hygiene reasons.

5.6   Shelter

The construction of shelter, along with the ability to start and
control fire, is considered to be one of the essential elements
for the colonisation of cold environments. Noble and Davidson
(1997) claim that the archaeological evidence for the construction of shelters prior to the emergence of modern humans is
"more like wishful thinking". As with the evidence for fire and
burials, the evidence for these purported shelters can be easily
explained by the work of natural geological processes. The
mechanics of erosion and accumulation of flood materials can
produce what appear to be the remains of artificial man-made
structures. The earliest structured shelters are found on the
Ukrainian plain dated to at most 30,000 years. These constructions, made from mammoth bones and mandibles, are the first
evidence of deliberate planning, and for Noble and Davidson
conclude that the capacity for building artificial shelters did
not belong to our pre-human hominid ancestors.

There is no doubt at all that early hominids were using caves as
shelters, although Binford and Ho (1985) challenge the conven-
tion that to identify a site as a habitation is to automatically

assume "at home" behaviour. Their interpretation of "The Cave
Home of Beijing Man" challenges many assumptions about the
"lifeways" of *H. erectus*. The picture of early man living in a
cave, making fires, and bringing hunted animals back to a home
base for consumption, is problematic. There is no evidence
for campfires around which these hominids sat, slept, cooked
or made tools. Bone accumulations in caves indicate that *H.
erectus* was a scavenger rather than a hunter, so the image
of Beijing Man sitting around a fire eating hunted foods is
a myth. Binford and Ho argue that *H. erectus* life was not
"culture-based", but more like 'conditioned behavior'.

5.7   Environmental management

One of the quintessential hallmarks of human behaviour is
our ability to manipulate the environment and to exploit just
about every available resource. In most parts of Africa we find
materials associated with early modern humans, which reveal a
system of acquisition that requires organization, planning and
cooperation. However, this type of organization is not found
in sites associated with our presumed hominid ancestors, *H.
erectus*.

It appears that the first modern human population, that emerged
in Africa, remained very low for some time, judging by the types
of faunal assemblage found in caves that date to 90,000 years
ago. Klein (2000) has pointed out that wherever we find large
ungulate and tortoise bones together with the larger specimens
of marine shells, we can infer a low population density that is
not over-exploiting the available resources. Tortoises require
eight to twelve years to reach adulthood, and edible shellfish
take two to five years to mature. A small population is what
we should expect with a scenario of the sudden emergence of a
new species.

We don't know why, during approximately 60,000 years of apparently successful expansion throughout Africa, *H. sapiens*

managed to become the only species in Africa. Archaeological evidence reveals that some of the first modern humans in Africa had the ability to exploit many new resources, which McBrearty and Brooks (2000) believe may been one of the main reasons for population growth due to the apparent increased survivorship of human infants. They stress the importance of access to marine foods for the human diet. Fish and other marine foods are rich in Omega-3 fatty acids, which are beneficial for development of the human brain. In Southern Africa we also find evidence of controlled burning, which stimulates corm[11] production and adds a plant food rich in carbohydrate to the diet. In addition we find exploitation of marine mammals such as seals and birds, including penguins. Moderns humans seem to have also engaged in exchange networks of raw materials and made deliberate forays into far-away regions to extract or mine exotic materials. McBrearty and Brooks have shown, through geochemical analyses, that certain materials had their source 320 km from an excavated shelter. We certainly don't need to ask the question why *H. sapiens* may have moved out of Africa, and eventually reach all other continents on Earth. After all, *H. erectus*, like many animals, were able to migrate successfully, no doubt, tracking suitable environments. Nevertheless, we cannot deny that *H. sapiens* must have had something special in their sustainability 'toolkit' in order to become the only hominid in existence during a rapid migration program over a very brief period of time.

Stiner et al. (1999) have analysed several sites with food remains around the Mediterranean and have found that from 200 ka up until 44 ka, the pre-modern hominid populations were exceptionally small and highly dispersed. Modern humans overlapped with Neandertals at this time, which suggests that it was not population pressure or competition for resources that caused modern humans to wipe out the Neandertals in a short period of time. We do however know that there was a mini ice age that brought drier and colder conditions to many areas of

---

[11]Fleshy edible underground bulb-like stem emerging with new growth.

Europe around 40,500 years ago. Mellars (2006) suggests that this may have delivered the "*coup de grace*" to the Neandertals, most likely because modern humans were better equipped, both culturally and technologically, to deal with this dramatic climate change. Wherever we find deposits of shells and bones that have been damaged by fire, tool marks and percussion fractures, we can infer that modern human (*H. sapiens*) activity is involved. Tortoises and mollusks, being practically sessile, are the easiest prey to catch and do not require an advanced technology. These easily caught animals are the most sensitive to hunting pressures and overharvesting. The size of these animal remains can be seen to decrease gradually throughout this time, but the earlier deposits contain quite large specimens indicating exceptionally small initial population densities. Diminution of tortoise remains can be seen at around 44ka indicating an increase of predation, which Stiner et al. believe to indicate population pressure on resources. Here we see a totally different settlement dynamics to both the European Neandertals and other *H. erectus*.

5.8   Summary

I have assimilated arguments from many scholars in the field of archaeology who argue that it is entirely likely that the ancestors of *H. sapiens* were not evolving as a 'trend' toward becoming 'human'. Early hominids, as argued in chapter 4, were merely bipedal primates, who left no mark on the landscape to indicate that their behaviour was any different to extant Great Apes. After 4-5 million years of 'evolution', we see the emergence of crude stone tools, which, on closer examination, are little advanced in complexity than those that extant chimpanzees can produce, with a little encouragement. I have suggested that these first simple single-strike stone tolls may have been an accidental 'discovery', rather than the product of an 'inventive' mind.

Approximately 1.8 mya, we find a new stone tool in the 'tool-kit', which at first sight, appears to imply greater sophistication of manufacture. This so-named 'hand-axe' sometimes appears to have a more symmetrical shape, which many scholars believe entailed a greater level of sophistication of production. The recruitment of 'mental templates', together with a greater understanding of cause and effect, are also claimed to have been necessary for its maker, our immediate ancestor, *H. erectus*. However, the correlation of this stone tool, which is prolific in the archaeological record, with an advance in cognition has been challenged by many scholars. It has been shown that the symmetry of this tool was more likely the result of the reduction possibilities of the raw material of the rock used (Toth and Schick, 1993; Bickerton, 2002). It has also been noted that the 'hand-axe' did not change in shape or complexity over a period of one millions years, and a geological spread of 10,000 miles (Bickerton, 2002). We find little evidence of human-like cognition or any ability to move technology forward over both time and distance.

It has often been claimed that our hominid ancestors were gradually evolving more complex social systems, and this may have led to the evolution of language for communication (e.g., Dennett, 1995; de Ruiter and Levinson, 2008; Pinker, 1994; Schoenemann, 2006; Ulbaek, 1998). However, a robust re-evaluation of the archaeological record by many skeptical scholars (e.g., Binford and Ho, 1985; Gargett, 2000; Noble and Davidson, 1997) has led to the conclusion that our hominid ancestors show little, if any, advance over the social organization of extant Great Apes. The lack of evidence for ritual burial (Gargett, 2000), together with the grizzly evidence for cannibalism within species (Fernández-Jalvo et al., 1999), speaks for a more ape-like social system, if that's what we could call it.

Stahl (1989) rightly cautions that "the image of our early ancestors as social creatures cooperating around a cozy hearth, sharing the profits of a day's hunting, seems less tenable in the face of more rigorous analyses of site formation". It makes sense that the use of fire for warmth, cooking, and as protection against predators, would have greatly enhanced the survivorship of hominids (Bellomo, 1994; James, 1989; Wuethrich, 1998). The placement of hearths in or near cave sites, implies the use of home bases, and perhaps the maintenance of social groupings. However, the archaeological record does not fit with this conception of life for pre-human hominid ancestors. There is no evidence for modification of sites to be used as home-bases (Ranov et al., 1995), and many of the claimed hearths can be equally accounted for by the work of natural processes.

If pre-sapiens hominids were building shelters, making campfires, burying their dead and engaging in other social rituals, it seems fair to ask the question why the archaeological landscape is not peppered with evidence for these behaviours. Claims for an association of pre-humans with an active ability to recognize the causal nature of the physical world has been treated with skepticism and has been rigorously challenged by many scholars in the field. Likewise, claims for the use of ritual or symbols by our hominid ancestors have been soundly discredited. We can conclude that most of the behaviours that we recognize as uniquely 'human' came to be only with the sudden appearance of *H. sapiens*.

6. Genetics and Development

6.1 Introduction

The purpose of this chapter is to gather support from genetics to further underpin the paleontological and archaeological evidence that modern humans (*Homo sapiens*) evolved 'suddenly' and fairly recently. In chapter 4, I put forward the paleontological evidence suggesting that hominids evolved in sudden leaps, possibly numbering only two major transitions, leading to anatomically modern humans. Although hominid anatomy evolved in only two stages, I argue that all of the cognitive traits that we usually assign to human uniqueness arose suddenly in just one crucial step with the emergence of *H. sapiens*.

The fossil record in Africa closely matches the latest genetic evidence for the recent emergence of *H. sapiens* (Adcock et al., 2001; Drayna, 2005; Mitteroecker et al., 2004; Pearson, 2004). Of course, a change in anatomy does not necessarily equate with a change in cognitive ability. However, with the arrival of *H. sapiens*, it seems that changes in the architecture of the brain have emerged along with the change in anatomical features of the skull. Research in neurogenesis confirms the radical differences in the building of the human brain during development (Ramus, 2006). Hints of modern cognition, as technology and ritual behaviour become visible in the archaeological landscape, are found soon after modern humans are found in the fossil record.

The Recent Emergence of *Homo sapiens*

Molecular data reveals that humans are very closely related
to each other, with mtDNA and the X and Y chromosomes
in all modern humans showing very low diversity, especially
compared to extant apes (Mitteroecker et al., 2004; Pearson,
2004). Wallace (2005) notes that because of a critical bottleneck in our early history, our overall genetic diversity is less
than that found within and between contemporary chimpanzee
subgroups. It seems that this bottleneck coincides with the
sudden appearance of anatomically modern humans.

There is overwhelming evidence pointing to a common genetic
origin, which can be traced to a founder population of hominids
that lived around 120,000 years ago. It happens, that from a
morphological viewpoint, the first appearance of anatomically
modern humans can be found in Africa, and this emergence
is also dated to around 120,000 years ago (Stringer, 2003).
Recent work with founder mutations confirms that all humans
are members of a single family, bound together by our shared
genome (Drayna, 2005).

Current consensus is that an anatomically modern population
emerged in Africa, and subsequently replaced all other hominid
populations living there at that time. Pearson (2004) suggests
that these first modern humans grew in population over a
period of 60,000 years, but remained in Africa, explaining the
greater genetic diversity that is found there today. After 60,000
years, *H. sapiens* dispersed to the Near East, South East Asia
and Australia, and then on to Europe, East Asia, and the
Americas.

Crow (2002) believes that a saltational change in a single male
may have selectively isolated humans from other hominids. He
has identified a block of sequences within Xq21.3 that trans-
posed to the Y chromosome short arm, and this mutation is
thought to have produced a speciation event that led to the
sudden emergence of humans. A mutation arising in a single

male can spread rapidly through sexual selection within a small group, which would have isolated itself from the main group. Crow also believes that this single mutation has led to the brain asymmetries related to language.

Modern human cognition, together with all of the support systems required for language, emerged at this crucial time in our evolution. Recent work in neurogenesis attests to the fact that a single change in the timing of development of the human brain has built a radically different brain architecture, which has laid the foundation for this cognitive revolution (Elliott, 2001; Gilissen and Simmons, 2001; Hayakawa et al., 2005; Rakic, 1995; Ramus, 2006; Schoenemann et al., 2005; Schwartz and Maresca, 2006).

6.2 Relationships - Great Apes and Humans

Current fossil evidence suggests that the first split from a Great Ape ancestor occurred around 5-7 million years ago, although Arnason et al. (2000) place the split at up to 13 million years ago. In this section, I show that it is not feasible that the important cognitive differences between our Great Ape ancestor (using extant Great Apes as proxies) and ourselves can be accounted for by the gradual accumulation of point mutations in the genome.

Most of the changes in the anatomy of the Great Apes are due to postnatal changes in gross shape components. Early hominids appear to have followed the developmental pattern for the other Great Apes, that is, changes that occur postnatally. Genetic difference between the Great Apes, according to Mitteroecker et al. (2004), mainly exert their effects during the later stages of ontogeny, in contrast to the genetic differences in humans that affect early ontogeny. It is apparent that the genetic differences that are present in the early ontogeny of humans gives rise to the large morphological differences from the (other) Great Apes. As detailed in chapters 3 and 4, small genetic changes in homeobox genes during embryonic and postnatal development, can have profound effects on resultant phenotypes.

Although it is widely claimed that our genome is 98.5% similar to that of the chimpanzee, Ramus (2006) points out that human and chimpanzee genomes may differ by 20% in terms of functionally significant differences. He finds that only about 30% of proteins are identical while 70% of proteins differ in amino acids. A detailed comparison of chromosome 21 reveals a staggering 83% of protein coding differences. In addition, there are larger-scale duplications of sequences that have occurred in our *recent* evolution, which Ramus believes to be a major factor in our phenotypic differences. He adds that about 70% of our gene variants are human specific, and a large proportion are expressed most highly in the brain. The claim that human and chimpanzee genomes differ by only 1.5% is for Ramus incorrect and misleading.

Schwartz and Maresca (2006) concur with Ramus. They believe that molecular systematics has placed too much weight on single changes in protein building genes rather than the replicatory duplications that have occurred in the structural genes that govern development and determine many of the major changes in morphology. There is a major contradiction between genetic similarity and morphological dissimilarity between humans and chimpanzees. These major morphological differences, Schwartz and Maresca believe, can be understood in terms of changes in the genes that regulate development, rather than in minor changes in the genes that build proteins. We should also remember that chimpanzees have 48 chromosomes (2n=48 karyotype), whereas humans have 46 chromosomes (2n=46 karyotype). The fusion of the chimpanzee chromosome 12 and 13 to form the human chromosome 2, had the potential to cause radical changes in the development of the hominid phenotype, and more than likely was the cause of eventual reproductive isolation.

A case in point is the evolution of the hominid pelvis and femur that facilitated bipedalism. Lovejoy et al. (1999) point out there are "no genes for bones", but a highly complex interaction of transcription factors that influence the differentiation pathways of the cells involved. A simple change in a signalling pathway during development has radically altered the morphology of the pelvis (along with the femur), although the cells that build the tissues (the raw materials) making up these structures are highly conserved. Dramatic anatomical changes resulting from pattern formation shifts during development have caused sudden changes in anatomy. This change in the pelvis and femur does not stem from natural selection acting on its variation and "guiding" its evolution. The HoxD sequence, that controls the formation of the vertebral column, pelvis, and limbs, resides on the human chromosome 2. This reduction in the number of chromosomes from the pongid 48 to the human 46 is believed by Bowers (2006) to be the quintessential mutation that led to bipedality.

The important point to make is that, just as there are "no genes for bones", likewise, there are no genes for behaviour. We certainly have evidence for the underlying causes of the changes in anatomy from Ape to very early hominids and australopithecine, followed by the emergence of *H. erectus*. However, we have meagre evidence for significant changes in cognition within our hominid ancestors. In the following sections, I show that this is not surprising, due to the fact that brain growth, brain architecture, and consequent behavioural options, are usually highly conserved in mammals.

Ancestral Hominid Genetics

In this section, I present further evidence for the sudden emergence of the unique traits that are attributable only to *H. sapiens*. Arnason et al. (2000) point out that the replacement hypothesis does not fit with the theory for human evolution that posits a gradual change in hominid predecessors throughout different places in the world, essentially wherever *H. erectus* roamed. Rather, they argue that modern humans arose in Africa and replaced all other hominid populations, both in Africa and all other accessible continents. The limited amount of variation, in both mtDNA and the Y-chromosomes of recent modern humans, are further evidence of a recent emergence of a single population, with little genetic exchange with other hominids. DNA analysis of 50,000 year old Neandertal bones reveals that this species split from the human ancestral line approximately 600,000 years ago (Green et al., 2008). Around 500,000 years ago, a new wave of hominids[1] arrived in Europe from Africa, carrying with them the Acheulian handaxe, a stone tool that culturally defines the later Neandertals (Toth and Schick, 1993).

In 2008, analysis of DNA found no evidence of interbreeding between the modern humans and Neandertals when modern humans spread from Africa into Eurasia around 60,000 years ago (Green et al., 2008). This analysis was based on differences in mitochondrial DNA. However, a recent comparison of the DNA of some Neandertal and Denisovan fossils has overturned this finding. It appears that limited breeding between male Denisovans and Modern Human females took place as evidenced by DNA sequences found in some populations of Australians and Papua New Guineans (Cooper and Stringer, 2013). Also, Green et al. (2010) have since discovered that a very small portion of Neandertal DNA is shared by some modern human populations in Europe, but not in Africa. This suggests a certain amount of interbreeding as modern humans moved out of Africa and through western Eurasia. Neandertals became extinct shortly after the arrival of modern humans in Europe[2].

---

[1] *H. heidelbergensis* arose in Africa around 600,000 years ago (Rightmire, 1996), and is thought to be the possible ancestor of modern humans in Africa and the Neandertals in Europe.

[2] Excavations of the Les Rois cave site in France, dated to approximately 30,000 years, shows clear evidence that modern humans were interacting with Neandertals, but cutmarks on the bones of juvenile Neandertals indicate to Rozzi et al. (2009) that humans may have seen Neandertals

6.3   Neurogenesis

D'Arcy Wentworth Thompson published his classic work *On Growth and Form* in 1917 and his enlarged version in 1942. Thompson showed that simple scaling laws can change different structural forms like leaves, shells, skulls, horns, or fish through straightforward alterations in the timing of developmental events.

Elliott (2001) thinks that it is inevitable that evolution would exploit the variations that can be produced during temporal developmental events. He suggests that D'Arcy Wentworth Thompson would not be surprised by the findings of modern developmental biology, that evolution has tinkered with brain structures by altering the timing of neurogenesis. Naturally, as Elliott cautions, this is not the whole story as we must also take into consideration the huge amount of postnatal neuronal growth, which is greatly affected by the individual experiences of the organism.

Conservation of Neural Development

Finlay et al. (2001) ask the question, "when we speak of brain evolution, what exactly do we imagine to be evolving?". Due to the conserved nature of neurogenesis in mammals, we should be clear whether we view the brain as evolving as a whole or in a modular fashion. Their studies reveal that special selection of particular areas can occur, but these modifications play a minor role when compared with the covariance of all brain structure during development. The olfactory bulb appears

---

mainly as an edible resource. Rozzi et al. believe that this find may throw light on how the two species could have interacted, and this interaction most probably contributed to the demise of the Neandertals. It is entirely possible that modern humans (*H. sapiens*) did not recognize other hominids, including other *H. erectus* species, as part of their own *Homo* genera, due to radical behavioural differences.

to be labile in development, but all other brain structures grow or shrink together in evolution. Developmental processes place major architectural constraints on brain evolution. They point out that brain size appears to increase with body size at a predictable exponential rate, but the basic mechanisms of action, memory, communication, and cognition appear not to scale to brain size. Both the hummingbird with less than one gram of brain matter and the baleen whale with 5,000 grams do not appear to show much difference in complex behaviour. They both sing, defend territories and mates, raise their young, and migrate over long distances. Consequently, they find it unsurprising that it is difficult to link variation in a particular brain structure to individual species behaviour. Constraints on what sort of brain can evolve are evident when we look at cephalopods (octopus, squid, and cuttlefish). Gilissen and Simmons (2001) point out that this order of animals do not produce myelin to insulate their axons, so much space is taken up in their brains by large axons. In addition, due to their green blood, they can only carry about one quarter of the level of oxygen compared with red-blooded vertebrates. This oxygen-carrying capacity of their vascular system severely limits the amount of high-energy consuming nervous tissue that they can sustain.

6.4   The Human Brain - Escaping Constraints

Although there appear to be major constraints during brain growth, the occasional removal of constraints allows for major evolutionary changes (Gilissen and Simmons, 2001). A single parameter can have an enormous impact on brain evolution and may, in fact, produce an entirely different creature. There is ample evidence that a simple genetic change in the endocrine system, that builds neural tissue, can have a profound effect on the final architecture of the brain. McKinney (1991) has pointed out the importance of growth hormones in the mechanisms for building tissues. Two growth factors, IGF-1 and IGF-2, which are strongly correlated, appear to have been instrumental in the phenotypic variations we find in all vertebrates. For example, the variation of breeds that we see in the single dog species are attributed to the levels of circulating IGF-1. We find a similar picture within the primate order. Gorillas grow three times as fast as chimpanzees in order to attain their roughly threefold greater adult body size. IGF-2, the close correlate of IGF-1, acts early in fetal development and can increase mitosis in both brain and body.

Rakic (1995) contends that the mutation of a regulatory gene(s) that controls the timing and ratio of cell divisions creating the initial proliferative cells (stem cells) before neurogenesis begins, has most likely been responsible for establishing new patterns of connectivity and an expanded cortical plate in the brain. Corticogenesis can be divided into two phases. The first phase produces the founder cells, and the duration of this phase determines the number of radial units in the cortex of a given species. The duration of the second phase regulates the number of neurons within these radial units (neurogenesis). A delay in the onset of the second phase by just a few days allows for three to four extra rounds of mitotic division, thereby producing a massive increase in the number of founder cells, which in turn would produce 8-16 times larger the number of columns and the expanded cortical surface. The species-specific size of the cortex is determined by the pool of proliferative cells in the early stages of embryonic development. Rakic concludes that a small modification of a regulatory gene appears to have played a major role in the evolution of the neo-cortex in the human brain. This has given us our expanded cortical plate and the enhanced patterns of connectivity to phylogenetically older structures.

In line with Rakic's findings, Finlay et al. (2001) have found that structures with neurons that are born late, grow proportionally larger as absolute brain size increases. Large isocortices are simply the result of being produced last in brain development. As the duration of embryogenesis is extended, the precursor pool of cells proliferate at a higher rate in this region. The human brain with its enlarged isocortex, Finlay et al. conclude, is likely to be a by-product of structural constraints later adapted for various behaviours.

The human brain shows a distinct pattern of gene expression relative to non-human primates, according to Cáceres et al. (2003). They find that approximately 90% of the genes that are involved in building the primate brain are more highly expressed in humans. Comparison of human and chimpanzee hearts and livers reveals no such differences. It is important to note that most differences between the two species result not from a difference in gene sequences, but primarily from alterations in their expression, and this has provided the basis for extensive modifications of cerebral physiology and function in humans. They believe that the higher levels of neuronal activity that occur in the human brain has important consequences in cognitive and behavioural capacities. Cáceres et al. also make the interesting point that these biochemical changes in human neural cells enable them to function longer than those in other primates. Gene-regulation changes support high levels of cerebral activity over the lot longer life span in humans.

Muotri and Gage (2006) point out the extraordinary diversity that is found within the neuronal population of the mammalian central nervous system, which may number as many of 10,000 types, each differing in molecular details. This provides cells with various excitation thresholds and distinctive firing patterns. The key to human cognition, they believe, may lie in the vast increase in connectivity afforded by the major increase in neurons, and the heterogeneity of the brain.

6.4.1   The Importance of White Matter

Schenker et al. (2005) point to the importance of both the cortex and white matter, which appear to form discrete neural systems recruited for cognitive functions and information processing, due to interconnectivity within the various brain regions. Barton (2001) notes that there appears to be some developmental conservatism in mammalian brain structures, with the scaling of neocortex size in primates to be nearly linear. He claims that the amount of grey matter in the human brain is what we should expect for a brain of our size. He notes that it is the white matter that has increased disproportionately in the human brain, and is responsible for the expansion of the total neocortex size. Zhang and Sejnowski (2000) have performed a theoretical analysis of the relationship between gray matter and white matter and find that as the gray matter increases in mammalian brains, the white matter increases disproportionally faster, according to a power law. They believe that the volume of white matter in relation to gray matter is highly conserved in the mammalian class, but admit that the brain has many components, and it is logically inconsistent to claim that all of these components would scale to the same power laws.

Hayakawa et al. (2005) have found an important difference in brain microglia between humans and chimpanzees. Possibly due to changes in regulatory sequences, the Siglec-11 gene is prominently expressed in the human brain. This gene is thought to have arisen as the result of a gene duplication and recombination event. They make the point that this presumed gene duplication, which is universal to modern humans, is a recent gene conversion event. Microglia have traditionally been thought of as the immune cells of the brain, springing into action after brain trauma. However, Tremblay et al. (2010) have found that microglia also play an important part in the modification of synaptic structures. Their observations suggest that the dynamic activity of microglia are crucial to circuit remodeling and brain plasticity, contributing to learning and memory in the healthy brain.

Further evidence for a simple change in the human brain is provided by the comparative analysis of primate brains. The human frontal lobes are not, as a whole, larger than expected for an ape of our size, but the prefrontal region is disproportionately large in humans (Rilling and Seligman, 2002). Generally, temporal lobe volume scales with overall brain size in primates, but their analysis shows that the human temporal lobes have increased in volume dramatically, mainly due to a significantly larger amount of white matter. We see the beginning of this pattern with chimpanzees, which are commonly accepted as our closest hominoid ancestor. The disproportionate size of the human temporal lobe white matter augments the number of connections linking temporal and prefrontal cortex (Rilling and Seligman, 2002). The primate temporal lobe contains functional subdivisions for visual processing, auditory processing, social behaviour, and association cortex, all implicitly involved in the production and comprehension of human language.

Schoenemann et al. (2005) concur with Rilling and Seligman. They have extensively utilized magnetic resonance imaging brain scans of 11 primate species in order to determine where the human brain differs from the expected brain architecture for a primate of our size. They stress that the human brain is not simply a scaled up version of a primate brain. In fact, although the human brain is around three times larger than expected for body size, some areas are actually smaller than we should expect. The primary motor and premotor areas occupy a smaller proportion of the cortex in humans than in other primates. However, when it comes to the prefrontal area of the human brain, we find radical changes. Their high-resolution MRI scans have excellent gray matter/white matter differentiation, which allows for the measurement of white matter volumes in the primate brains scanned in their study. They find that average values for total volume and white matter volume were greater than expected for a primate of our size, but not the gray matter. Although there appears to be variation among human brains, they found that the proportion of white matter volume can be as much as around 80% larger than expected. Schenker et al. (2005) point out that white matter is composed predominately of axons, and not only do we find a greater number within a larger brain, but also their width increases. Axonal width is important in that it allows for more rapid transmission of signals across larger distances in the brain.

This expansion of white matter, particularly in the anterior portion of Broca's area, has implications for human language, especially semantic information. Not only does Broca's area show an expansion of white matter, recent experiments by Amunts et al. (2010) have revealed a much more complex functional diversity than has been previously reported. A simple classification of Broca's areas 44 and 45 into two cytoarchitectonic areas is no longer enough. They have found differences between cortical areas that reveal different receptor types, which play a crucial role in neurotransmission and signal processing properties. They believe that the identification of these new regions in Broca's provide a new structural basis for the organization of language in the brain.

Schoenemann et al. (2005) note that the greater interconnectivity between the prefrontal cortex and many other brain areas has been emphasized in discussions of language evolution. The dramatic changes have allowed for an increase of processing temporal information, which in turn has given humans an understanding of causality. An understanding of causality, they believe, is based on the ability to remember the temporal order of past events, and understanding causal relationships underpins general-problem solving abilities, which was critical for the development of elaborate technology.

Human Brain Metabolism

The human brain has an unusually high metabolism, which Cáceres et al. (2003) believe is supported by expression changes in glial cells, which in turn support the energy requirements for neurons. The human prefrontal cortex has expanded over 200% compared to apes. This growth depends on the proliferation of supporting glial cells, which is due to an over expression of homeotic genes leading to a higher allocation of embryonic stem cells to the brain, with fewer stem cells allocated to the body (McKinney, 2002). The human body grows slower than other primates, and it seems, other hominids. Each stage in human development regresses in onset time, namely with birth, puberty and death. Life history parameters have had important implications for the social structure of humans in comparison to other primates and hominids.

Recent work by Rae et al. (2003) emphasizes the importance of the glutamate/glutamine cycle in the brain enabling a fast, energy efficient recycling of neurotransmitter glutamate. Astrocytes, which make up most of the white matter in the brain, are an important and necessary support for neuronal activity. Astrocytes are not only involved in the metabolic support for neurons, but have been shown to play an important information processing role in the central nervous system (Araque, 2008). Astrocytes, like neurons, are able to sense synaptic activity as an input signal, and are able to integrate these signals into an output signal, thereby actively modulating neuronal excitability and synaptic transmission.

Gold (1995) has also shown how increases in circulating glucose concentrations have very broad and robust influences on brain functions, enhancing learning and memory. Hormonal responses to certain situations can release glucose stores into the circulation and influence the laying down of memories. For example, stress hormones released by a particularly arousing experience will ensure the storage of the incident in memory. Learning and memory seem to be greatly enhanced in humans due to the increase in glucose metabolism, which influences the energy production of several neurotransmitters synthesized from that glucose.

It is reasonable to conclude that the recent evolutionary change in the amount of white matter in the human brain is implicated in the sudden behavioural changes that we find in the archaeological record around 100,000 years ago. Not only has this increase in white matter allowed for the metabolic support of a huge increase in non-volatile memory, it has also provided enhanced patterns of connectivity to phylogenetically older structures.

6.4.2 Language Genes?

If language is a uniquely human trait, and as claimed by many, arose through gradual change[3] leading to a complex universal grammar, then we should expect to find a genetic underpinning that subserves language. However, finding a correlation between the genotype and phenotype for language, as Fisher (2006) points out, has not only been elusive, but also has led to overly simplistic and abstract views of the nature of the gene. He reminds us that genes do not specify behaviours, but build complex biological systems by making regulatory factors, signalling molecules, receptors and enzymes. To understand the genetic influences on human cognition, he believes, we should understand the inherent molecular complexity of developmental systems. We must embrace the idea that the genome assembles complex arrays of molecular machines, which vary between species. Most importantly for Fisher is the discovery of how the genome controls the way in which a gene is expressed (turned on and off) in time and space.

Fisher (2006) finds it just as ludicrous to talk of a "language gene" as it is to talk of a "smart gene" or a "gay gene". While not denying that variability in cognitive and personality traits may have a genetic cause, a simple mapping of genotype to phenotype is "merely an illusion". He is critical of studies that introduce abstract notions of neural circuits that seldom reflect the underlying genetic architecture. Phenotypic studies generally parse cognitive processes for ease of experiment, but this parsing cannot link directly with individual sets of neural activity.

---

[3]In chapter 7, I survey many theories purporting to claim not only what would have been precursors to language, but also how language is thought to have evolved gradually after the split from our pongid ancestor.

In the early 1990's a regulatory gene *FOXP2* was discovered and named the "grammar gene", due to its implication in the language disorder of a particular large three-generational family. Later work did actually find a single nucleotide change in a copy of *FOXP2* on chromosome 7, and this appeared to be responsible for the affected members of this family. *FOXP2* is one of 40 different types of forkhead protein, which play an important role in controlling genetic cascades during development, and as Fisher highlights, are especially important during embryonic development of the central nervous system. *FOXP2* is a regulatory gene. It switches on and off the expression of other genes by coding for a type of regulatory protein called a transcription factor. A slight alteration in the DNA interferes with the ability to regulate its target genes properly, and appears to be associated with reduced levels of functional *FOXP2* protein in the brain. This disruption appears to be enough to lead to an impairment in speech and language. Friederici (2006) has shown that the *FOXP2* is correlated with the development of motor-related brain circuits, and its involvement may be functionally related to speech pathology, but, like Fisher, is adamant that it cannot be considered "the language gene".

FOXP2 is not unique to humans. In fact, Fisher points out that this protein was present in the common nonlinguistic ancestor of humans and mice that lived over 70 million years ago. *FOXP2* acts as a 'master switch', which regulates patterns of gene expression during development of the lung, heart and gut, and is important for the development of neural tissue in both humans and mice. He thinks it highly unlikely that this gene evolved to subserve language, as it is highly conserved in other mammals, which are clearly nonlinguistic.

Interestingly, the differences in the *FOXP2* protein between humans and chimpanzees give an indication that these changes arose within the last 200,000 years. So it seems to Fisher, that this gene, although implicated in the functioning of normal language and speech ability, did not evolve gradually to support language. *FOXP2* plays a role in the improved motor sequencing skills in humans as well as other aspects of cognition, such as procedural learning, or the unconscious acquisition of skills through practice. This supports the suggestion that alterations of the *FOXP2* protein may be involved with impairment in many aspects of central nervous system function, which causes problems in multiple brain regions. Fisher has found that it is especially apparent in the disruption of procedural learning, and seems to produce deficits in language output skills, mainly difficulty coordinating movements of the throat and tongue.

FOXP2 is just one of a number of regulatory genes. Normal eye development can be disrupted by a mutation in another forkhead gene, *FOXC1*, due to a insufficient production of transcription factor. However, we do not say that this is the "eye gene", nor should we say that *FOXP2* is the "language gene" (Fisher, 2006). Moreover, *FOXP2* is not within the human DNA *for* language and speech, and has not gradually evolved to provide humans with these abilities.

6.5 Summary

There is no doubt that genetics has a major role to play in the study of evolutionary development. Our genetic make-up has provided important information for ascertaining when and where modern humans emerged (120,000 years ago, in Africa). It has challenged previous assumptions about the pattern of human evolution, where it was thought that each of the current human groups evolved gradually from separate hominid ancestors. Genetics has also supplied insights for how the developmental 'program' plays out in conjunction with our genes.

Most importantly, genetics has shown how the 'gene for' notion offers little credibility. For example, their is no gene 'for language'. The adaptationist program has been seriously compromised by the now commonly held view that the complexity of all animal cognition is unlikely to be unraveled by searching for a gene *for* an individual trait. By examining our 'developmental program' instead, we can utilize the extraordinary amount of information from genetics that provides tantalizing insights into the evolutionary developmental programs of our hominid ancestors, and also us.

7. The Evolution of Language

7.1 Introduction

Language is just one of the cognitive traits that appear to be uniquely human, but the case for its evolution is the most difficult to argue. Brains do not fossilize, but even if they did, we would not be able to say for certain whether a particular brain had the capacity for language. At the present time, we cannot even tell, just by looking at the brain of a modern human, if it contains a language faculty.

There are several aspects to consider when dealing with the evolution of language:

1. Did language evolve *for* communication?

2. Did language evolve *for* other reasons (organizing thought, planning, reasoning about causes etc.)?

3. If language evolved suddenly, together with the emergence of *Homo sapiens*, what enabled this faculty?

The main claim in this chapter is that language did not evolve *for* communication. In fact, language did not evolve *for* anything. Due to the paucity of evidence for any precursors of language, 'naturally' spoken or signed, in any other primates, and the lack of evidence for any advance from this situation in our hominid ancestors, I argue that it is highly unlikely that

189

language 'evolved' gradually and incrementally from existing primate cognitive abilities.

In chapter 5, I argued that our hominid ancestors appear to have been incapable of the cognitive behaviours that are purported to be the foundational requirements for a language faculty to operate. Although the archaeological record does not speak for itself, I believe that much of the behavioural complexity attributed to pre-human hominids has been vastly overstated.

For around the first 4 million years of hominid 'evolution', there was absolutely no footprint on the archaeological landscape. There were no stone tools or any other artefacts, no burials, in fact not one iota of evidence that hominids had made any cognitive advance over their Great Ape ancestors. There is no reason to assume that any form of a language faculty had emerged within these hominid populations.

Around 2.6 million years ago, simple, single strike stone tools arrive in the landscape, and around 1.1 million years later, a more 'worked' stone tool emerged. Many scholars have argued that this new stone tool reflected a breakthrough in hominid cognition, involving planning, and the ability to hold a mental template for a desired finished product. I have challenged this assumption in chapter 5, and will revisit this issue in this chapter.

In the following sections, I present, but mostly refute, some of the arguments put forward for a gradual evolution of human language and the associated traits that are thought to have arisen to support the language faculty. I argue that not one of the underlying mechanisms that we posit as necessary to support the language faculty lends itself to an adaptationist explanation. I also argue that language, along with many of its support systems, arose as a discontinuous evolutionary novelty, along with the sudden emergence of *H. sapiens*. The main point of difference to my argument is the claim that the saltational evolution of *H. sapiens* heralded the total package of modern human cognition, and the capacity for language.

Due to the fact that there are no surviving hominids other than modern humans (*Homo sapiens*), many scholars have turned their attention to the study of chimpanzees in order to establish some sort of precursor/s for the language faculty. Those taking the gradualist Darwinian approach, involving incremental change, are usually the most enthusiastic about finding precursors for language in other primates (e.g., Cheney and Seyfarth, 2005; Dennett, 1998; de Ruiter and Levinson, 2008; Pinker, 1994).

Schwartz and Maresca (2006) put this enthusiasm down to the assumption by many scholars (usually neo-Darwinian), that because humans share 99.4% of their DNA with chimpanzees, they must be closely related in cognitive traits. However, Schwartz and Maresca warn of the error of this simplistic assumption as most of the genetic material that humans share with chimpanzees represents primitive retention, and is phylogenetically meaningless. Accordingly, molecular systematics has placed too much weight on single changes in protein building genes, rather than exploring the replicatory duplications that have occurred in the gene regulatory sequences that govern development, and have determined many of the major changes in morphology during evolution. They note the major contradiction between genetic similarity and morphological dissimilarity between humans and chimpanzees. These major morphological differences can only be understood in terms of changes in the genes that regulate development, rather than in minor changes in the genes that build proteins.

Ramus (2006) concurs with Schwartz and Maresca by pointing out that human and chimpanzee genomes may differ by 20% in terms of functionally significant differences due to larger-scale duplications of sequences that have occurred in our recent evolution. Many of our gene variants are human specific, and importantly, a large proportion is expressed most highly in the brain. The claim that human and chimpanzee genomes differ by only about 1.5% is for Ramus incorrect and misleading.

We should not forget that chimpanzees have also evolved in

the last 6-7 million years or so since the split from our last common ancestor, and in similar ecological conditions. This point should be taken into account when we mistakenly posit external selection pressures 'pushing' only one group of primates toward evolving a language faculty. Moreover, under a Darwinian scheme of constant improvement of species, we would need to view extant chimpanzees as much more cognitively advanced than the common ancestor of both humans and chimpanzees. So, this common ancestor, that lived 6-7 million years ago, would have had even less cognitive sophistication than extant chimpanzees. Of course, if we avoid the Darwinian gradualist 'trap', we have no need to contemplate what sort of cognition this ancestor possessed!

Although I reject Darwin's theory of gradual evolution, the most important part of his theory, that evolution proceeds by descent with modification, stands. It therefore seems a reasonable enterprise to explore the cognitive abilities of chimpanzees in order to gain some insight into how a language faculty may have emerged. Unfortunately, as argued later in this chapter, this enterprise has yielded few encouraging results.

This chapter carries the underlying theme that we should look to the evolution of structure before worrying about what function that structure now serves. This is in direct contrast to the adaptationist approach, that typically discerns the current function of a structure in order to determine which selection pressures might have *caused* the structure to gradually evolve.

7.2   Language *for* Communication

The argument that language evolved *for* communication is usually associated with the utility of E-languages. E-languages are public languages used by certain populations in order to make 'external' communication possible. They are epiphenomenal objects that arise from the triggering effects of exposure to a particular community of speakers (Chomsky, 2005). According to Chomsky, E-languages, due to their ephemeral nature, are not appropriate objects for scientific study in terms of their evolution, which is purely historical. In fact, Chomsky believes that we should not even use the term 'evolution' when we speak about cultural artefacts belonging to humans – more specifically, E-languages.

Along similar lines, Mendíl-Giró (2006) is concerned with the notion of evolution of language as a social object. He questions the analogy of the evolution of language with Darwinian gradual change leading to improvement, with the elimination of undesirable elements. This incongruent notion of languages (as E-languages) having evolved as adaptive systems for better communication, leads to untenable assumptions about linguistic change. Rather than focusing on historical changes in E-language and how they have evolved, he argues that we should concentrate on the evolution of I-language[1], which can be thought of as a linguistic species, with each member's language organ, or phenotype, built by both the human genotype and developmental processes. For Chomsky (1980), I-language is something every normal human acquires unconsciously during early development and is essentially a product of the mind/brain. The universal aspects of I-language (the language faculty) are the common endowments built by the human genotype and appear to be unique to *Homo sapiens*.

Regardless of the focus of study (I-language or E-language), implicit in the idea that language evolved *for* communication is the basic assumption that language evolved gradually and incrementally, and was 'honed' by natural selection. Anderson and Lightfoot (2002) believe that arguments that follow the gradualist approach for the evolution of language, that is, from the simple to the complex, can be seen as left-overs from

---

[1]Chomsky (1980) coined the term 'I-language', meaning internal, intensional, and individual, in order to claim what he believes to be the only really valid (scientific) study of human language.

nineteenth century thinking, where languages were treated as external objects and evolved law-like, with directionality.

7.2.1   Gradual Evolution of Language *for* Communication

Many theories of language evolution are built on the assumption that language evolved in a classical Darwinian fashion, that is, as an adaptation honed by natural selection in order to improve the fitness of those with the ability to communicate using speech or gesture. Some scholars argue that language emerged soon after the split of the hominid clade from our primate ancestor around 5 million years ago, while others believe that it developed with the evolution of *Homo* about 2.6 million years ago. Very few scholars believe that language developed later in hominids, with the evolution of anatomically modern *Homo sapiens* humans. It is difficult to find an account for the evolution of language that does not include a fundamental assumption that language evolved *for* communication, and, of course, incrementally and gradually. Even Bickerton (2005b), who does not strictly subscribe[2] to the gradual and incremental evolution of language, believes that *something* must have kick-started language, and the obvious candidate for him was the pressure to communicate.

Bickerton (2002) argues that, even if gradualism in evolution is accepted, we cannot apply this theory to the evolution of language. Language did not evolve seamlessly out of animal communication systems. He believes that "there is simply not one scintilla of evidence: simply a blind faith that, if evolution is gradual, and we are where we are, we must have got here, far as it may seem, in a series of incremental steps".

The crux of the selectionist account, as Botha (2002) points out, is that any trait exhibiting complexity must have evolved

---

[2]Although Bickerton envisages a beneficial spiral as language and thought co-evolved, a viewpoint that is difficult to disentangle from a gradual and adaptational approach.

through natural selection. The problem for Botha is that much work in language evolution lacks a precise definition of the entity/entities that have evolved. Thus he has difficulty in finding any plausible accounts in the literature that attempt to explain how either the language faculty *or* any of the myriad of human languages could have evolved incrementally[3].

Jenkins highlights the problem for a selectionist account for the evolution of 'parts' of the language faculty.

> Suppose, thirty years ago, when long movement was standardly assumed in linguistic work, we had asked why language is designed such that it has long movement. Consider now the following possible answers: long movement facilitates reproduction, winning friends and influencing people, communication, gossiping, ease of processing, or perhaps long movement was favored by natural selection. We can easily see why such answers are useless as explanations for language evolution. For, today, having learned that there is short movement, not long movement, we ask again why language was so designed. The nonanswers are the same: gossiping, winning friends, etc. (Jenkins, 2000)

Most selectionist accounts for the evolution of language merely state that language would have been adaptive *once* it had already evolved. This logical error still persists in much of the literature on the evolution of functionally useful traits, although this mistake was pointed out many years ago by Piattelli-Palmarini.

> It is one thing to assess the current utility of a biological trait, and quite another to explain the origins of this trait in terms of its current utility.

---

[3]So far, Jackendoff (1999) appears to be the only scholar who has attempted to account for the incremental evolution of language from the simple to the complex (see section 7.6).

> Even if a trait *is* useful and actually enhances the
> life expectancy of the individuals who possess it, this
> fact does *not* grant the inference that the trait is
> there *because* it is useful (Piattelli-Palmarini, 1989).

Piattelli-Palmarini believes it more likely that useful traits
have been exapted for novel purposes, rather than having been
modified or adapted to serve that trait.

Reid (2007) has made a similar point that "to perceive a phenomenon as adaptive is not to explain it". He adds that a
selectionist or adaptive account of the evolution of a useful trait
will usually find an advantage under *current* ecological conditions, but this does not explain its supposed adaptive advantage
at the point of origin. In fact, Reid makes the important point
that the emergence of novel properties would often obscure
the evolutionary history of a particular trait. In other words,
self-organization can lead to complexity, but the factors that
go into forming this complexity will often not be found in the
lower levels of an evolved trait.

Serious criticism also comes from Rosselló and Martín (2006),
who complain about adaptationist stories for the evolution
of language that lack testability, and often the complete lack
of relevant linguistic properties to be explained. To rigidly
follow the "ultra-Darwinism" line of adaptation and gradualism
they believe, leads to a certain dogmatic adherence to theories
that should be updated in the light of new evidence from all
disciplines involved in the evolution of human cognition and
language.

Nevertheless, many other scholars are quite firm in their belief
that a Darwinian approach involving selection and gradual
change is the right way to proceed. For example,

> acceptable speculation has been greatly narrowed by
> the recognition that any account of language origins
> must be consistent with the principles of evolution
> by natural selection (Noble, 2000).

Accordingly, for Noble, all of the intermediate stages of the evolution of a language capacity must have had an adaptive advantage and "gradual development is more plausible than catastrophic change". Schoenemann (2006) also claims that language *must* have origins that "are substantially older than the appearance of anatomically modern *H. sapiens*". He believes that Darwin was correct in assuming that the development of the brain and the elaboration of the vocal cords were the *result* of the evolution of language.

Probably the boldest claim is made by de Ruiter and Levinson (2008), who, despite overwhelming evidence to the contrary, believe that our nearest cousins, the chimpanzees, are not only highly intelligent, but are able to master most human tasks that are not mediated by language. They believe that communicative intelligence has evolved slowly over the six million years of separation from our primate relatives, and has eventually led to the evolution of spoken language. This of course raises the question why it is that only hominids were able to convert this communicative ability into language. This notion is a case in point for the observation that selectionist stories imply that "humanity required initiative and effort, and the apes remained apes because they didn't exert themselves enough" (Leakey, 1992).

Pinker claims that,

> [t]hough we know few details about how the language instinct evolved, there is no reason to doubt that the principle explanation is the same as for any other complex instinct or organ, Darwin's theory of natural selection (Pinker, 1994).

His basic claim is that language evolved in order to enhance communication, which in turn allowed for a greater fitness for planning, sharing memories, and perhaps gossiping to maintain social relationships. He further argues that the human vocal tract is "tailored to the demands of speech", and human auditory perception shows specializations for decoding speech sounds into linguistic segments (Pinker and Bloom, 1990). However, as noted by Jenkins (2000), language may help to find mates, win friends and influence people, but it tells us nothing about the structural properties that give rise to language. Further, the auditory perception mechanisms and the human vocal tract *constrain* the phonetic system and the possible articulations of language; they were not "tailored" to the *demands* of speech (Piattelli-Palmarini, 1989; Jenkins, 2000). Even babbling in infants, vocal or manual, is a brain-based expressive activity (Petitto and Marentette, 1991). It is not a phenomenon linked to the maturation of the vocal apparatus, as some language recapitulationists would have it.

Also following the Darwinian gradualist tradition, Dennett (1998) is a self-proclaimed "unblushing adaptationist". He cannot understand that anyone could believe, on the one hand, that something like the language organ is innate, but on the other hand, that this did not mean that it was a product of natural selection (Dennett, 1995). In answer to Dennett's bewilderment, Chomsky simply argues that we can concoct as many adaptationist stories as we like, but "it remains to explain how the biological endowment developed; the problem is simply displaced, not solved" (Chomsky, 1988).

Falling back to 'special pleading', Dennett believes that language plays an important role in structuring the *human* mind, but adds that "the mind of a creature lacking language – and having really no need for language – should not be supposed to be structured in these ways" (Dennett, 1991). Elsewhere, he claims that vervets evolved a 'necessity' of communication and expression of emotion because of the stressful lives they lead (Dennett, 1998). Apparently, their lifestyle is so simple, and their communication 'needs' too few, so they could not make 'use of' the features of human language. Thus, he adds, "we can be quite sure that evolution hasn't provided them with one". A similar spurious argument is offered by Milo and Quiatt (1993), who think that language, with a capacity for abstraction, "would not have been essential to pre-*Homo* hominids" as pre-modern hominid cultures do not appear to be highly structured or differentiated by group.

Many scholars feel uneasy with the idea that the capacity for language could have arisen by any other means than traditional Darwinian gradual evolution. Falk believes that the fossil record indicates that "language's manufacture has also been of long and proud duration", comparing it with "Noble machines like the Rolls", which are not invented overnight. She adds that "although it is true that the cultural products of earlier *Homo* are dowdy by comparison, a sudden appearance of something as universal and complex as language *does not feel right*" (Falk, 1992, my italics).

In the final section of this chapter, I argue that although the sudden emergence of human cognition and language "does not feel right" to many scholars, we have little alternative but to accept the fact that this is a more plausible theory than that of emergence through gradual "manufacture".

Along with the assumption that language evolved gradually *for* communication, there is an underlying notion that human language evolved from animal calls, which are often considered to be communicative. In section 7.4, I present evidence that refutes the idea that animal calls are 'intentionally' communicative.

In the following section, I survey the many theories that aim to identify other precursors for language, apart from animal calls. Most gradualist accounts of language evolution recognize that many of what we might call subsidiary systems are needed to support a generative communicative system. These possible support systems can be classified as,

- Symbolic Systems

- A Rich Theory of Mind

- Social Intelligence

- Pragmatics

- Knowledge of Causality

- Enhanced Memory

I will argue that we have no plausible theories linking precursors for language in any of our primate 'cousins', which includes all of our hominid ancestors.

7.2.2   Symbolic systems

For many scholars in evolutionary linguistics, the most important break-through in the evolution of language must have been the ability to think in symbols. However, Piattelli-Palmarini et al. (2008), although recognizing that the use of symbols, or words, to refer to objects or events in the world not immediately in our presence, is an important part of language, think that many scholars have overrated this link. They make the point that words in 'human' languages are not simple one-to-one relations to the world. Accordingly, we often overlook the importance of the complexity of single lexical items in language, a point I will return to in section 7.5, relating to proto-language. Nevertheless, if we are hoping to find an initial break-through of symbol use in our hominid ancestors, then we have to first see if there is a hint of this capability in other primates.

There is overwhelming evidence that suggests that other primates do not use symbolic communication in the wild. Snowdon (1993) maintains that the many studies of chimpanzees and gorillas in the wild reveal a lack of complex vocalizations. He also points to the disappointing results with experiments raising infant chimpanzees and humans in the same environment. After seven years of training, one chimpanzee could produce only four words, even though the chimpanzee vocal tract is actually capable of producing quite a few phonemes of human language. Moreover, Bickerton (2007) points out that animal calls are indexical and not symbolic. They have diametrically opposed qualities in that animal calls are meaningless or misleading in the absence of an immediate referent, whereas symbols can refer to anything in the human experience whether present or not, or even if the referent is non-existent in the 'real' world. Cangelosi (2000) believes that apes have great difficulties in learning symbolic relationships due to the fact that their brains, particularly in the prefrontal cortex area, are structured differently, and therefore do not provide the functionality to support this ability. Apes can only learn symbolic relationships under certain experimental conditions that take into account the pragmatic aspects of communication. He finds that apes merely emulate human language without comprehending its symbolic representational function, which makes it very difficult to be able to extend or generalize this learned ability to novel situations.

Tomasello et al. (2005) have found that the ability to collaborate, and the motivation to share symbolic artifacts, is the key difference between humans and other species. Tattersall (2006) notes that symbolic behaviour arises naturally in human infants as they engage in pretend-play, where an object can stand-in for another object or concept. At this point, human infants start to use words as symbols. Terrace, (in Lieberman et al. (1991)), points out that human infants use single words to communicate about objects for the sheer joy of communicating, and without expecting a reward. After many years studying chimpanzee behaviour, he finds that they only use 'words' in a game with the expectation of a pleasurable reward. They do not use words to spontaneously communicate about things, and cannot be trained to use any syntactic rules.

The debate over whether animals use symbolic communication systems continues with much zest. However, Provine (2005) urges caution when attributing conscious control over seemingly intelligent and complex behaviour in other animals. We cannot help trying to rationalize the behaviour of other animals, including sometimes other humans, in terms dictated by our "own inner voice". Bolhuis and Wynne (2009) make a similar point. They have shown how during experiments with other animals, we cannot help anthropomorphizing animal behaviour in terms of human cognitive patterns of behaviour. Savage-Rumbaugh (1994) believes that chimpanzees do not learn words or symbols, but learn only to connect certain behaviours with the selection of a symbol. Work with the chimpanzee (bonobo) Kanzi has shown that he will select a sign purely on the basis of what he anticipates will happen afterward, like getting a banana. She thinks that knowing how to use a symbol to get someone to give you a banana is not the same as knowing that "banana" represents a banana. It is just understanding that a symbol will generate a certain pattern of behaviour, but only if that resultant behaviour has previously been experienced on a regular basis. If the behaviour (getting a banana) is not fulfilled, a chimpanzee will show confusion, which suggests that the chimpanzee does not have the concept of a banana, just the encoded 'behaviour' of the matching symbol selected.

Apart from the ability to use symbols, we also need a rich theory of mind for language to operate successfully. Sterelny (2003) stresses the fact that we rely on the symbolic meaning in each of our utterances to be understood by our audience. Despite many years of experimentation with Great Apes, there is little evidence that they have any rich ways of representing the minds of others.

In chapter 5, I argued that we have little reason to believe that the symbolic behaviour we associate with the unique cognitive traits of modern humans (*H. sapiens*) had made any inroads into the cognition of our hominid ancestors. It is often assumed that the making of stone tools heralded the emergence of symbolic behaviour in our hominid ancestors (e.g., Gergely and Csibra, 2005; Bridgeman, 2005). However, Chimpanzees and orangutans regularly make and use tools, but this certainly does not entail the use of any form of symbolic behaviour, including language.

Homo neanderthalensis, who is purported to have been the most cognitively advanced of our hominid ancestors (d'Errico et al., 2003), left behind little evidence of symbolic behaviour, according to Tattersall (2006). Like other *H. erectus* populations, they did not ritually bury their dead (Gargett, 2000; Noble and Davidson, 1997), they were not technologically 'inventive' (Gargett, 1993; Semaw, 2000; Ambrose, 2001; Ranov et al., 1995; Bickerton, 2002; Johnson, 1989; Chase, 1990), and they did not leave any artifacts that may have been of a symbolic nature. Chomsky (2005) supports the idea that the origin of language was more likely the result of a genetic event that rewired the brain in modern humans, causing the sudden and emergent explosion in expression of symbolic thought through symbolic behaviour, as revealed by the archaeological record.

7.2.3   A Theory of Mind

Having a 'theory of mind' allows one to gauge another's mental state by observing their behaviour. Cheney and Seyfarth (2005) suggest that because nonhuman primates lack a theory of mind (not being able to attribute mental states to other minds), they fail to recognise the causal relations between behaviour and beliefs. As a result, they are unable to map their thoughts onto a communicative system in any predictable way.

Povinelli and Vonk (2003) are very cautious about attributing a theory of mind to any animal apart from humans. Humans undoubtedly have mental states together with the core ability to think about mental states. Humans also understand that other humans have mental states that are probably similar in nature. In addition, humans have the ability to realize that one's own mental content may be different from that of another human. However, attributing the same abilities to chimpanzees is something that the human mind simply cannot help doing. That is, human minds can't help "distorting the chimpanzee's mind, obligatorily recreating it in its own image". They accept that chimpanzees form concepts, but argue that these are related to statistical regularities in behaviour. For example, chimpanzees would build a concept of 'threat display' after having experienced what follows (being hit) from seeing another chimpanzee pursing his lips or noticing hair bristling. The reason we often liken chimpanzee minds to our own is that we are so closely related in evolutionary terms, and often appear to have similar behaviours. Accordingly, our folk psychology interprets chimpanzee behaviour in terms of assuming that chimpanzees understand what must be going on in other chimpanzees' minds. Povinelli and Vonk believe that most anecdotal evidence of chimpanzee behaviour involving reasoning about what each other see, want, know and believe, are constructions within the human's mind, and not anything to do with what is going on in the chimpanzee mind. A chimpanzee may appear to be manipulating another chimpanzee, but this behaviour can result from the ability of the chimpanzee to represent individual links between particular behaviours and responses, which have been learned from past experience.

Our seemingly rich 'theory of mind', and our ability to contemplate complex abstract notions, must play a significant role in language processing, according to Marcus (2006). We also need a rich theory of mind for language to be understood by one another. Sterelny (2003) has argued that when it comes to theories of the evolution of language, there is no point in using analogies with apes. Although experimental work with apes may provide a base-line, we cannot argue for 'real' language without grounding it within a rich theory of mind. Povinelli and Vonk suggest that the idea of 'mental states' may well turn out to be an "oddity of our species' way of understanding the social world", and they remain skeptical about anecdotal evidence for a chimpanzee's ability to represent the mental states of others.

An Intentional Mind

Understanding the intentions of others is foundational for deciding what the other is *aiming* to show. Intentional action, according to Tomasello et al. (2005), is based on three components tied together by control-system principles. Many animals exhibit intentional behaviour based on these principles. Animals have goals, capacities to attempt the attainment of these goals, and monitoring systems that regulate their enactment, taking into consideration changes in the environment. Most animals have feedback mechanisms based on their perceptual systems that allow them to monitor the environment and adjust their course of action.

Tomasello et al. explain that goals can be separated into external and internal. For example, if an agent wants to open a closed box, then the opened box would be considered to be the external goal. The internal goal of this experience is the person's mental representation of the desired state of the opened box. This internal entity is the goal that determines the action plan of the agent. An intention includes the internal goal and the action plan that is needed to achieve that goal. Animals use the current mental map of reality, as well as their stored knowledge base/skills, in order to effectively pursue the internal goal. In some animals, an emotional reaction accompanies a disappointment at failure or happiness at success, or even surprise at an accident. They note that many goals may contain many subgoals that all contribute to decision making.

Many animals have this level of intentional behaviour. Tomasello et al. are more concerned with how they understand the intentional actions of others. Do other animals perceive and understand that an actor has goals and behaves in a manner to reach those goals? An observer has to not only recognise that another actor has a goal, but also to understand and perhaps predict the processes involved in the actor's persistence in achieving that goal. This level of understanding of the intentional action of others assumes that an observer knows that the other animal can actually see and rationalize what an appropriate behaviour might be in the circumstances. It also allows an observer to predict what another animal might do, even in novel circumstances. Tomasello (2000b) believes that nonhuman primates do not recognise each other as intentional agents. His claim is based on the fact that in the wild, chimpanzees:

- do not point or gesture to outside objects for others.

- do not hold objects up to show them to others.

- do not try to bring others to locations so that they can observe things there.

- do not actively offer objects to other individuals by holding them out.

- do not intentionally teach things to others.

The ability to collaborate, and the motivation to share symbolic artifacts, are for Tomasello et al. (2005) the key differences between humans and other species. They observe that the earliest types of dyadic engagement in human infants involves mutual interaction directly with other humans. This interaction is mainly observed through emotional expressions and behavioural turn taking, and is "the glue" that holds protolanguage together. Mutual gazing involving the capacity to share emotions appears to be a universal feature among all human cultures.

Experiments undertaken by Tomasello et al. show that human infants by about 6 months old understand that other humans are animate objects, and spontaneously produce behaviour, often with predictability, in familiar circumstances. By 9 months, infants understand emotional intentions, like whether an adult is teasing or acting in good-faith. At 12 months old, infants understand pointing and may even display dissatisfaction if an adult does not acknowledge the infant's attempt to share the experience, by attending to the object and commenting on it. By 15 months old, many infants can recognize another's goal and even imitate the necessary actions that would be required to complete another's desired result. This type of understanding in human infants allows for imitative behaviour leading to a powerful form of cultural learning. Once infants have learned that they share the same intentional behaviours as others, they start to understand and engage in collaborative exercises. Tomasello et al. point out that this engagement with others is quite different from social interaction in general, the type we find in other animals. Each participant understands both roles of engagement and can engage in role reversal, even helping the other with their part if needed. An infant may even take over a recalcitrant adult's role in the activity. By this time, infants also have begun to engage in linguistic communication. Linguistic symbols are involved in collaborative engagements, allowing shared perspectives on any number of topics[4].

Despite their general agreement with Tomasello et al., Brownell et al. (2005) believe that the motivation for human infants to share their emotions, desires and intentions with others may be grounded in their attachment relationships with carers, and may be limited to this interaction. Their experiments reveal that human infants show little motivation to share this sort of collaborative behaviour with peers, and it does not emerge until the close of the second year of life, or sometimes well into the third year. The theory that attachment relationships underlie initial interactions between infants and carers is also supported by Charman (2005), whose studies of children with autism shows that the cognitive and affective components of human intentional systems are highly intertwined. Neuropsychological systems that subserve the motivation or capacity for shared intentionality may be impaired in children with autism, who generally fail to enter fully into human culture.

If attachment relationships do underpin the affective compo-

---

[4]Tomasello et al. (2005) believe that the atypical development of children with autism, who do not engage or interact with other persons, may offer a clue to both the phylogeny and the ontogeny of human social cognition.

nents of an intentional mind, as Brownell et al. and Charman believe, then it is difficult to support the idea that our hominid ancestors had made any advance in this area over that of chimpanzees. Although attachment relationships are evident between mothers and their infants in most social animals, this is viewed by Archer (1998) as purely instinctive behaviour. He has shown that there is a crucial difference in behaviour between humans and chimpanzees when the loss of an infant occurs. Initially, chimpanzee mothers show anxiety and distress, even hostility, when separated from their infants, just like human mothers. However, chimpanzees soon loose interest in their offspring if they die, either through natural causes or attack by other individuals in the group. Both random and systematic killing, including cannibalism, of infants and juveniles in wild chimpanzee populations in Gombe National Park, has been well documented (Wilson et al., 2004). Infanticide and cannibalism among male chimpanzees has been witnessed by Hamai et al., and in one case, a mother and daughter 'team' was not adverse to killing and eating their own infants, as well as others in their group. In chapter 5, I presented several cases of similar behaviour among our hominid ancestors. Cannibalism of infants is interpreted by Fernández-Jalvo et al. (1999) as a lack of empathy of hominids for their own species. He suggest that, due to the similar butchering techniques, breakage patterns to extract the marrow, and identical pattern of discard, these hominids viewed other hominids of their own species as food, along with any other mammal. We also have a complete lack of ritual disposal of bodies in any of our hominid ancestors (Gargett, 2000). We can only conclude that our hominid ancestors were not engaging in the same sort of emotional interaction as that which is a 'normal' component of the human emotional and social development that partly forms an intentional mind.

Great Ape Intentional Behaviour

Tomasello et al. (2005) offer evidence that nonhuman primates appear to understand each other as animate agents who produce behaviour spontaneously. They point out though that much controversy surrounds the notion that apes understand goal-directed behaviour.

An alternative view considers that apes understand only behaviour, and not another's mental state behind that behaviour. For example, Povinelli and Barth (2005) ask whether we can discern a meaningful difference between representing mental states versus representing behaviour. They believe that Tomasello et al.'s interpretation of apes' ability to represent the goals and intentions of other apes may be overstated.

It appears to Povinelli and Barth that experiments purporting to show intentional behaviour in apes actually presuppose that apes can distinguish between various actions of others in terms of a variation between mental states. They argue that just because apes are able to form complex representations of another's *behaviour*, this does not mean that apes actually have complex representations of intentional states.

A similar, but more radical stance is taken by Provine (2005) who is skeptical about how much the role of intention plays in both animal *and* human behaviour. He questions our "tendency to presume rational, conscious control over processes that may be unconscious and not require a ghost in the neurological machinery". Accordingly, we often overestimate the amount of conscious control and "are misled by an inner voice that generates a reasonable but often fallacious narrative and explanations of our actions, and we use this account to interpret the actions of others". He points out that seemingly intelligent and complex social behaviour in bees, ants and termites is most likely achieved with little, if any, conscious control, and he urges a more conservative stance when explaining the actions of animals, including many of our own.

Tomasello et al. (2005) counter the claim that chimpanzees might be just reacting to behavioural cues. They argue that apes have the ability to understand intentional action in terms of goals and perceptions. This view is based on experiments that they believe show that chimpanzees appear to recognize when a human is unable, or just unwilling, to give them a piece of food. The chimpanzees apparently will gesture and persist in trying to extract the food from an unwilling human, but will give up if they perceive that the human is just unable to give them the food item. The chimpanzees, they argue, seem to understand whether a human is well-meaning, or otherwise, in their behaviour, a similar reaction observed in 9- and 12-month old infants. Tomasello et al. also believe that their experiments show that chimpanzees understand that others can follow gaze direction and this affects their behaviour. When a dominant can see food, and a subordinate recognizes this fact, the subordinate will invariably not attempt to access that food. However, if the subordinate knew that the dominant could not see the food, it would access that food. This implies some sort of rationally driven behaviour.

These results are, however, disputed by Povinelli and Barth (2005), who argue that although certain behaviours are undoubtedly underpinned by different intentions, any animal does not need to understand another's behaviour in terms of sharing complex representations, but rather just needs to simply keep track of invariant behavioural traits. For example, an ape may decide not to approach food if it is in the unobstructed path of a dominant ape. Based on previous experience, according to Povinelli and Barth, this is a wise choice.

Despite the claim that chimpanzees seem to show a very basic ability to understand a conspecific's expected behaviour, Tomasello et al. (2005) point out that apes do not appear to have even the most basic motivation to share psychological states with others. They show little desire to share goals or perceptions with others. Apes do not point, or use any other signals to declare or inform another, nor do they offer things or engage in any form of negotiation. Apes do not offer help to one another, nor do they ever smile at one another. Some apes have been trained to point for humans but show no understanding when humans point for them (Tomasello et al., 1997). In fact apes continue to point in some inappropriate circumstances, like when an experimenter has their eyes closed or has a bucket over their head. Apes simply do not understand the factors necessary for the sharing of this communicative behaviour.

Due to the fact that apes seem only to demonstrate an understanding of intentional behaviour in a competitive environment, Tomasello et al. (2005) propose that a Machiavellian account of cognitive evolution may be the best way to proceed. This account, they believe, would offer an explanation of the selective processes that may have operated in hominid evolution to produce what now seems to be the innate skill of intention-reading that emerges early in human development. The motivation to understand others' intentions in a competitive environment may have provided the "pathway leading to participation in collaborative cultural practices". This may be a plausible suggestion, but as argued later, the evidence for 'collaborative cultural practices' appears to enter the scene only with the arrival of modern humans (*H. sapiens*).

7.2.4 Cooperation and Collaboration

Many experiments with primates have been devised in order to determine their capacities for cooperative behaviour. The ability, or even the motivation, to collaborate or cooperate with a member of one's own species, is one of the traits that is essential for the sharing of a communicative system like E-language.

Tomasello et al. (2003) point out that chimpanzees basically never really share anything. They never indicate to each other a source of food that they could have solely for themselves, and only in special circumstances do they share. Most food is obtained in a competitive context. They note that even language-trained chimpanzees seldom, if ever, use their newly

acquired symbols to share information with others coopera-
tively or declaratively. This motivation is, of course, a necessary
prerequisite for the sharing of language, which is impossible
without shared intentionality. Falk (1992) points out that chimpanzees use their "hard-won" language skills almost exclusively
for immediate gratification, mostly to obtain food from their
experimenters.

More than 50 years ago, experiments claimed to show that two
juvenile chimpanzees cooperated to pull a heavy box with food
on top. Tomasello et al. (2005) reveal that it was made quite
clear in the original experiment that these two chimpanzees were
each trained separately to pull when the trainer said "Pull!".
When the chimpanzees were put together and told to "Pull!",
it looked as if they were engaging in cooperative behaviour in
order to haul the box. In short, chimpanzees simply do not
understand the nature of collaborative structure.

Tomasello et al. (2005) elaborate on another experiment where
one person hides food in one of several opaque buckets and
another person approaches the bucket to tilt it over so that a
chimpanzee can see that the food is in this particular bucket.
When the buckets are offered to the chimpanzee, the chimpanzee
of course remembers where the food is and takes it. However, if
the second person, rather than tipping the bucket over, instead
points to the bucket where the food is and uses gaze direction
between the chimpanzee and the bucket, then the chimpanzee
chooses randomly. Chimpanzees do not understand what pointing or purposeful gazing means. Human infants understand this
game at one year of age, but chimpanzees never understand the
collaborative context of the game.

Studies by Butterworth and Franco (1993) show that by the
age of 6 months, a human infant will turn its gaze toward the
direction of another's gaze, as long as the target is within the
infant's visual space. By 12 months, human infants understand
the pointing gesture. They will look into the direction of
where another is gazing and pointing. So, it appears that by

twelve months old, human infants appear to have developed the cognitive ability to extrapolate an imaginary trajectory in visual space. By 14 months they will produce the pointing gestures themselves, and by 18 months, an infant will attend to another's gaze or pointing, even if the target is outside of their immediate visual field.

Chimpanzee Hunting Behaviour - Cooperation?

Rather than attempt to show cooperative behaviour 'in the lab', some have argued that we need to look at behaviour 'in the wild' to find signs of this trait (Schuster, 2005). Much has been made of chimpanzee hunting behaviour with two competing theories. Tomasello et al. (2005) argue that when it comes to hunting, or group defense from predators or other chimpanzee groups, chimpanzees do not act in concert or show any ability to coordinate plans. What may appear to be cooperative behaviour in a hunt is actually just every individual assessing the state of the hunt at any given moment and acting in its own interest.

Schuster challenges Tomasello et al.'s interpretation of chimpanzee hunting behaviour. He argues that other animals, like lions and hyenas, also indulge in cooperative hunting behaviour, implying that they share joint intentions for a shared outcome. Schuster believes that perhaps we notice shared intentional behaviour in humans more because it is often applied to activities for "which it was not designed". He thinks that some animals have evolved this ability for cooperation for a very limited set of activities, and these coordinated actions have become innate.

Tomasello et al. (2005) believe that the observations of so-called cooperative behaviour among chimpanzees can be explained more plausibly in other ways. Hunting chimpanzees do not have a shared goal, nor do they play a 'role' in the hunt. Schuster believes that the 'role' that a chimpanzee plays in the hunt determines the share of the food that they obtain, and also that chimpanzees may change roles depending on the current circumstances of the hunt, indicating that they have a capacity for role reversal and perspective taking. However, Tomasello et al. counter this argument by observing that age, hunting experience, and dominance factors, can equally explain the outcome for food sharing. In other words, different cognitive mechanisms may be at work, other than an ability to share intentional behaviour.

From competition to cooperation

For Ulbaek (1998), there must be a continuity from ape-like to human cognition. He is enthusiastic about tool-use and social systems that have evolved within chimpanzee groups in the wild, but finds that most of chimpanzee behaviour is quite selfish. It seems that chimpanzees are motivated only to engage in behaviours that will further their own needs and not the needs or wants of others. The key to the evolution of human cognition for Ulbaek, is reciprocal altruism, which apparently enabled fitness to our hominid ancestors. He feels that he has solved the often raised problem of how to account for the evolution of language, which many feel has evolved for communication: cooperative enterprises do not allow for individual competition. Therefore, altruism offers an edge that leads to greater fitness. Reciprocal altruism for information sharing may, according to Ulbaek, enhance fitness for an individual.

Tomasello et al. (2005) also suggest that hominid individuals or groups may have channeled their competitiveness into reciprocity and cultural conformity. They believe that a small biological change may have moved primate cognition onto some sort of cultural evolution that now makes a big difference in human cognition. This change, they consider, may simply involve a larger brain with more computing power or perhaps just a larger working memory. A larger working memory would allow hominids to hold more things in mind simultaneously. Tomasello et al. think that language derives from the same underlying cognitive skills that motivate humans to share and direct the attention of others.

The question arises of when primates began to 'channel' their natural competitiveness into cooperative reciprocity, or altruism, and cultural conformity. Unfortunately, we keep coming back to the same answer. The archaeological record is completely silent on the matter for pre-sapiens hominids. There is no evidence for cooperative enterprises like built environments or even home-bases, apart from unmodified cave sites (Tattersall, 2006). Due to the fact that we have not found any cultural artefacts with pre-sapiens hominids, it is difficult to imagine what sort of cultural conformity Tomasello et al. might have in mind. Pre-sapiens hominids did not bury their dead nor did they perform any sort of rituals for their dead (Gargett, 2000), a practice we recognize as a distinctly human cultural activity. The sobering evidence that points to a more chimpanzee-like social structure comes from many sites where cannibalism has been practiced by within-species hominids, not for ritual purposes, but for food procurement (Bermúdez de Castro et al., 1999; Fernández-Jalvo et al., 1999; Rozzi et al., 2009).

7.2.5   Social intelligence

Although the precursors for language are difficult to find in other animals, many researchers feel that it has not been a problem to find continuity of many behaviours from apes to humans. For example, Gibson and Ingold (1993) observe behavioural continuity with apes and humans, citing intra– and inter–group conflicts including lethal encounters along with close bonding with genetic relatives.

Social Organization

Pinker (1997) argues a special case for primate social organisation, which he believes promotes a need for better information exchange, and this *need* provided the pre-adaptation for hominids to move into a new "cognitive niche". In a similar vein, Cheney and Seyfarth (2005) believe that the earliest stages of language evolution may have grown out of the apparent primate ability to discern knowledge about social relations. This suggestion is grounded in experiments showing that nonhuman primates appear to be able to acquire information about their social companions via vocalizations, which are discretely coded and hierarchically structured. They are cautious about attributing the ability to form complex concepts to nonhuman primates, but nevertheless believe that the initial significant change in hominid cognition led, over several millions years, to modern human language. Theirs is a "gradual" approach with the demands of social life creating selection pressures for complex, abstract conceptual abilities. They suggest that

> the primate mind evolved in an environment characterized by intense social competition, that such competition created selective pressures favoring structured, hierarchical, rule-governed intelligence, and that such social intelligence shares many formal features with linguistic intelligence (Cheney and Seyfarth, 2005).

Tomasello et al. (2005) have shown that several primate species demonstrate that they have an understanding of third-party social relationships. Although many mammals recognise their own kin, and also form alliances and coalitions with other individuals within their group, only primates appear to have an understanding over and above their own social relationships with others. They appear to be able to formulate categories of social relationships and have the ability to track both their own current status within a relationship, as well as those that hold within other members of the group. Along similar lines,

Calvin and Bickerton (2000) believe that the development of reciprocal altruism based on grooming, together with basic call systems and gestures such as pointing, especially in a foraging context, may have been the precursors for the emergence of a protolanguage.

However, some scholars believe that we make too much of the apparent chimpanzee ability to track social relationships. Owren and Rendall (2001) note that most animals exhibit subtle anatomical variation in both size and shape of the supralaryngeal cavities, which allow an individual to produce vocalizations that mark its identity. Regular exposure to this variation of vocalization eventually leads to the ability to discriminate members within the group. Pika et al. (2005) have shown that most communication among chimpanzees can be shown to be ontogenetic ritualization. They have found that no social learning is involved in the gestures that they use in a social context. The study of several groups in the wild reveals that the picture is the same within groups and between groups. The inventory of behaviours, including pulling, punching, pushing, slapping and touching emerge naturally and without learning. Sexual rituals are also not learned, but follow instinctual routines.

Chimpanzee instinctual behavioural routines are in stark contrast to the gestural systems that most humans are capable of acquiring. Aronoff et al. (2008) point out that sign languages use the body and especially the hands to represent symbols for objects and actions. Nonhuman primates do not use their hands to represent an object other than their hands. Signers use a complex system of interaction between the body and the hands to express all aspects of their grammatical constructs.

An additional problem with the 'social organization' theory arises when we consider why chimpanzees, who have been under these same competitive social pressures for at least as many millions of years as hominids, have not developed (evolved) even a rudimentary level of language. Bickerton (2002) is critical of those who argue that *we* have achieved higher mental activities as a consequence of being immersed in a rich culture. He asks "how is it that I, but not chimpanzees, dolphins and hominids, have such a rich culture to be immersed in?". Deacon (1997) makes a similar point that humans can be ranked along with many other species as far as group size, brain size, social-sexual organisation, food requirements and the like, so we cannot single out special evolutionary pressures that would have applied just to one group of primates.

Arguing against special pleading for hominids, Frayer and Wolpoff (1993) make the point that many primates, birds, and other vertebrates exploit resources and exhibit complex seasonal behaviour without the 'need' for language. Hence, it is difficult to infer language skills from behaviour for any non-modern hominid. Along similar lines, Gargett (1993) notes that chimpanzees appear to have a structure that organizes and regulates access to food and mates without even a gestural language. He also reminds us that beavers transform their environment without language, culture, or technology. Due to the questionable nature of *H. erectus* 'culture', Gargett is justifiably skeptical of any arguments that premise language to be under strong selection in our ancestors. Even if we accept that the Great Apes have the cognitive ability to track relationships in a social context, this does not seem to easily or naturally translate into other contexts. Primates need hundreds or sometimes even thousands of trials to learn how to discriminate relational categories of physical objects (Owren and Rendall, 2001). It is not easy to see how the transition to human-like abilities to recognize categorical relationships between objects and events could have emerged without a major reorganization of cognitive architecture, a point I address at the end of this chapter.

A change in environmental conditions has often been cited as the impetus for a change in social behaviour in order to cope with new challenges. Calvin (2006) thinks that environmental fluctuations could have promoted the incremental accumulation of mental abilities that conferred greater behavioural flexibility in our hominid ancestors. A simple language, either gestural or vocal is inferred for *H. erectus* in order for this species of hominid to migrate out of Africa to Eurasia about 1.5 million years ago. This simple language, Milo and Quiatt (1993) speculate, may have included iconic gestures or minimal vocal units with no inherent semanticity. However, as argued later, it seems unlikely that even a very simple language for premodern *Homo* was in place. It is difficult to discern any different life strategies or social behaviour of *H. erectus*, over and above the behaviour of today's apes, that would 'require' language. Moreover, as noted in section 4.3.1, many primates (among other mammals) have, over millions of years, successfully migrated between continents without the 'need' for language. Gargett (1993) reiterates this point, noting that Miocene hominoids[5] migrated as far as *H. erectus*, presumably without the aid of language.

Milo and Quiatt (1993) nevertheless argue for a 'basic' form of communication that would have 'served' simple premodern hominid social relations, including group behaviour and sharing of resources and technology. They believe that advantages conferred by tool making would have evoked strong selection pressures for these hominids to acquire and exchange information.

However, as summarized in section 7.2.4, we have little evidence for an increase in social complexity or any advance in technology in any of our hominid ancestors. There is little (or no) evidence that points to differential social organization within any of the pre-human hominid sites found to date. Bickerton (2002) points out that the single item that represents any form of 'cultural' continuity related to *H. erectus* (the tear-drop shaped hand-axe) did not change in form from the time it emerged in Africa until we find it in Japan, a distance of 10,000 miles. Bickerton (2007) argues that what has been happening since the emergence of *H. sapiens* is cultural change, not something that is misleadingly described as "cultural evolution". We

---

[5]Global sea levels dropped dramatically around 16 million years ago, exposing land bridges and allowing for the migration of many apes from Europe to Africa, Eurasia, South Asia and China (Begun, 2006).

have precious little evidence for the same sort cultural change underpinned by an increase in social intelligence in any of our hominid ancestors.

From a developmental prospective, it may be the case that *H. erectus* brain growth precluded the acquisition of language. Tomasello (2000a) raises the possibility that up until three years of age, human infants have a different "use-based" language to that of adult humans and is based more on imitation of cultural norms. He argues that at this young age, children learn each verb one by one and their language is organized around these verbs. They rarely show creativity or productivity with their item-based lexicon. At approximately three years of age, most children reach a milestone in development when their particular grammar is "triggered" to produce more adult-like linguistic behaviour. As noted in section 7.6, *H. erectus* development and brain growth followed a more ape-like pattern in the early stages. Brain growth had reached 84% of adult brain size by one year old, a vastly different pattern of growth from human infants. The window of opportunity for language acquisition is likely to have been closed to *H. erectus* infants. Apes need a symbol system to be introduced to them by humans, in order to acquire a very rudimentary (if that) linguistic system. Likewise, human infants need to be provided with a symbol system in order to encode and manipulate propositional representations as well as indulge in analogical reasoning (Thompson, 2000). It is doubtful that adult *H. erectus* were biologically equipped to acquire a symbolic system that they could have imparted to the next generation of infants as a foundation for syntactic language.

Life-History

In chapter 4, I outlined the apparent life-history changes that occurred in our hominid ancestors caused by changes in developmental timing, and how they had radically altered their phenotypes. Comparative work by Blomquist (2009) illustrates this point. The change in growth patterns of hominids has implications for understanding the evolution of many of traits that underpin not only anatomy, but also those that may have caused a change in diet, cognition, social organization and life history. Blomquist points out that the entire human developmental sequence is generally delayed relative to that of chimpanzees, and this has had a major effect on human life-history events, especially the embryonic developmental trajectory of human cerebral hemispheres.

I have presented evidence for the fact that our earliest hominid ancestors had ape-sized brains and also appear to have had a similar life-history to extant Great Apes. A stasis of brain size lasted for approximately 5 million years after the split from our Great Ape ancestor. For this 5 million years of 'evolution', the archaeological record is completely silent, suggesting that our early hominid ancestors had not made any advance in cognition over that of extant Great Apes. Accordingly, the first hominids, including all of the australopithecines cannot be expected to have had a 'social' system that was in any significant way different from their Great Ape ancestor.

Our immediate ancestor, *Homo erectus*, had a larger brain than that of their australopithecine ancestor, but this increase can be put down to a dramatic increase in body size (Smith and Tompkins, 1995). *H. erectus* made stone tools, the first of which appear around 2.6 million years ago[6]. In chapter 5, I argued the case for evolutionary stasis of *H. erectus*. Their brain size remained unchanged, and their stone tools did not improve in any significant way throughout their 'evolution'. There is little (I argue no) evidence of symbolic behaviour that we associate with the unique cognitive traits of modern humans (*H. sapiens*).

---

[6]The first stone tools, known as the Oldowan tradition, are attributed to *Homo habilis*, although so far, fossils of this hominid have not been directly associated with finds of the first stone tools (Toth and Schick, 1993).

It has been pointed out by Johnson (1989) that in the Paleolithic, "the world was the foragers' oyster", so *H. erectus* had all of the rich resources across Eurasia from England to China, but this abundance did not appear to influence their ability to "organize the world in the same ways as modern *H. sapiens*". The ability to organize the world in order to escape population constraints has been due to the emergence of the modern brain of *H. sapiens*.

Bermúdez de Castro and Nicolás (1997) believe that our hominid ancestors more than likely had a life-history quite different from both chimpanzees and modern humans. Life expectancy is estimated to be only a maximum of 40 years. The fossil record of a group of 32 hominid individuals indicates an extremely low survivorship of infants and children together with a high mortality rate of adolescents and young adults. Mostly females died between 15 and 18 years of age and six other individuals died at a similar age, between 14 and 15 years. The high mortality rate of young females may explain the unusually small amount of infant remains due to the non-parental care of orphans. In addition, although the sex ratio for this population was 1:1, there appears to be a high degree of sexual dimorphism. Analysis of the mandibles (jaw bones) and dentition show a clear pattern of large size difference between males and females. This is a clear deviation from the regular pattern found in most primate populations, where large sexual dimorphism usually means a large departure from a 1:1 distribution of the sexes in a given community.

This evidence, together with the grizzly finds of within-group cannibalism among many hominid populations, tends to render implausible many of the theories claiming that our hominid ancestors were moving as a trend 'toward' modern human social organization. We should therefore be skeptical that the 'evolution' of culture, along with an increase in the complexity of social organization, was in any way involved in the evolution of human cognition and language.

7.2.6   Pragmatics

Kotchoubey (2005) reminds us of the importance of both pragmatics and prosody for the production of language. Pragmatics is such an important aspect of language, and, according to Kotchoubey, goes beyond semantics and expression. Pragmatics links the verbal and non-verbal components of language. The verbal component is used to communicate a goal, usually to direct someone to do something or to cooperate in achieving a goal. Non-verbal communication (prosody) is part of emotion itself, and gives language its "intonational colour and expressive power". While the left hemisphere in the human brain usually controls the verbal aspects of language, the right hemisphere processes affective prosodic information. Kotchoubey notes that affective prosody is "strikingly similar" in humans and other primates, and this similarity allows humans to easily identify the emotional content of other primate screams.

Similarly, Sperber and Wilson (1995) have found that human verbal communication is achieved by two different modes of operation. They believe that it is essential to incorporate all of the other non-linguistic elements into our models of how communication between humans is achieved. These elements include the time and place of the utterance, the identity and intentions of the speaker, and any other relevant properties that may influence the exchange. They find that modeling linguistic communication purely in terms of sentence coding often leads to many distortions and misperceptions of the data. We need to include all of our ostensive or inferential behaviour in our models of communication. For Sperber and Wilson, this approach does not lead to an intractable problem, as they believe that we can apply logical rules to our inferential behaviour. In other words, our conceptual representations may be amenable to logical processing[7].

---

[7] Although Piattelli-Palmarini (1989) is not enthusiastic about a formal system to prescribe the underlying factors relating to communicative efficiency, as he believes that there is little evidence of constraints dictated

Deacon (1997) observes that we share many traits with other animals, which appear to be innate and universally understood. He believes that subtle gestures and facial expressions may be similar to those used by other primates, but have nothing to do with language, although they may provide the subtle interpretations of intent that go along with language in humans. In her many years of working with chimpanzees, Savage-Rumbaugh (1994) has observed the uncanny likeness of chimp facial expression and gesturing with that of humans. She points out that we expect that human facial expressions are indications of how we feel emotionally, so we should expect that facial expressions on chimps should also reflect their inner mental states in the same way.

Although a science of the emotions has traditionally been seen as imbued with subjective prejudice in the past, Panksepp (1998) is sure that the neurophysical processes underlying our emotional feelings can be measured and manipulated in order to provide a realistic and scientific research program. Accordingly, he suggests that the study of animal emotions is a first approach to gain an understanding of the brain substrates that lay the foundations for understanding human emotions. Many of our motivations stem from hormonally regulated mechanisms deep within the limbic brain. Although these regions are common to all vertebrates, we see the expression of innate behaviours varying widely among species. Much of primate behaviour, including our own, can be traced to brain structures inherited from our reptilian brain. Ritualistic primate behaviour in showing aggression, courtship and greeting promote preservation of the species. Many of these nonverbal, species-typical forms of communication are important in the human communicative process.

Once again, we are faced with the problem of deciding at what point in the evolution of hominid cognition did the ability for

---

by the laws of pure logic. The processes of communication have little in common with formal logic and appear not to be linked to our natural ability for simple logic.

inferential behaviour arise. As noted in section 7.2.4, we have little evidence for intentional behaviour in the Great Apes. Their inventory of behaviours is the result of ontogenetic ritualization which involves no social learning, but follows instinctual routines (Pika et al., 2005). These behaviours emerge naturally in all groups of chimpanzees and seem to be of a completely different flavour from the ostensive behaviour of humans, which is not restricted to fixed repertoires, but can vary according to context.

For the first 5 million years of hominid 'evolution', we do not have any evidence for cognitive ability or any 'cultural' change that might be considered more complex that our Great Ape forebear. Our immediate ancestor, *H. erectus*, appears not to have engaged in any behaviour that we would think of as having emotional content, like ritual burial, adornment, or exchange tokens (Bickerton, 2002). Evidence of these human-like behaviours emerges only with *H. sapiens* around 100,000 years ago (McBrearty and Brooks, 2000).

7.2.7 Knowledge of Causality

Brownell et al. (2005) have shown that human infant interaction with others is normally a progressive development in which they acquire an ever increasing ability to share a reciprocal understanding of others' mental states. The growth in causal reasoning is believed to underlie a child's ability to explain and to predict the behaviour of others in an ever increasing complex social world. Markson and Diesendruck (2005) propose that children gradually develop a motivation to share their psychological states, and this may be driven by an ability to recognise causal, and thus an explanatory understanding of behaviour. This ability is not found in nonhuman primates nor in children with autism. This growing understanding of the causal and rational nature of behaviour, they argue, is the motivation underlying a sense of curiosity in children. They also believe that this ability is a "developmental primitive", due to the fact that it arises so effortlessly in human infants. By three years of age, children imitate intended acts by learning and testing, which shows that they apply a critical analysis of what should and should not be learned.

Watson (2005) accepts that the motivation for shared intentionality (see section 7.2.3) is an important starting point for the acquisition of language, but believes that a large body of evidence points to a different underlying capacity with which human infants are naturally endowed. This is the capacity to analyze the causal contingencies of a determinant world. Infants come to understand the world in terms of "all events are the effects of causal laws", and occur with a reliable probability over repeated instances. Infants come to know that the cause of variable behaviour lies in the actor or the situation. Young children have an innate ability to gather statistical evidence for the inference of goal-directed behaviour, and use this evidence to further make inferences for hidden physical causes.

The cognitive processes that allow the construction and prediction of causal events usually occurs late in the first year of human life. Lock (1993) believes that human cognition develops in parallel with the gradual structuring of the unique architecture of the human brain. He points out that cebus monkeys use tools successfully, but they do not show any sign of understanding of the causal processes involved in their operation. When it comes to chimpanzees, the story is the same. Tomasello et al. (2005) note that even when causal mechanisms seem to be reasonably clear, apes still do not show any understanding of the task at hand.

Schoenemann et al. (2005) point out that human brains have a considerable expansion of white matter, which affords greater connectivity, and has allowed for an increase of processing temporal information, which in turn has given humans an understanding of causality. A dramatic increase in white matter, especially in the anterior portion of Broca's area, has implications for human language, especially for semantic information.

They note that the greater interconnectivity between the prefrontal cortex and many other brain areas are important for language function. Oliver et al. (2000) have shown that we see a progressive specialization of cortical pathways that arise as the *product* of development and emerge in a relatively plastic manner through interactions with the intrauterine and postnatal environments. In general, they point out, the mammalian brain plan differs due to changes in developmental timing. The human brain shows a prolonged period of postnatal development, which is increasingly open to environmental influences and a greater exposure to worldly events. Friederici et al. (2006) have found two brain areas in the left frontal cortex that are involved in the processing of syntax. One of these areas of differentiation, Broca's (BA44/45), is a phylogenetically younger region and differs cytoarchitectonically with the layering of the cortex. Based on data from language-related imaging, they find that Broca's areas 44 and 45 are involved in hierarchical reordering of sentences due to a possible functional differentiation. Friederici et al. think it possible that BA44, which is present in the macaque, has just evolved to be bigger in the human brain. Together, BA44/45 have arisen as a crucial evolutionary trait that allows humans to process complex grammars involving structural hierarchies.

As noted in section 4.7.2, pre-sapiens hominids appear to have had a quite different developmental trajectory, especially for the brain. Hublin and Coqueugniot (2005) surmise that it is unlikely that *H. erectus* could have exhibited mental abilities similar to those of modern humans. Using CT scans of this fossil endocast, they determined that the *Homo erectus* infant had achieved 84% of adult brain size by one year old, whereas the human infant has only around 50% of adult size at this age. The developmental program for the *H. erectus* brain seems more chimpanzee-like than human. Due to the lack of technology associated with *H. erectus* and a complete lack of any artefacts for the australopithecines, it is unlikely that our ancestral hominids had any understanding of physical causality, which seems to be a major contributor to the faculty of language. This fact, according to Oliver et al. (2000), goes against the idea that the brain has become increasingly pre-wired over evolutionary time to recognize causal contingencies.

7.2.8   Enhanced Memory

An enhanced memory system has been advanced as a possible prerequisite for an increase in cognitive ability in our hominid ancestors. Tomasello et al. (2005) have posited the emergence of a larger working memory, which would allow hominids to hold more things in mind simultaneously. Vihman and DePaolis (2000) have shown that human infants up to 6 months old and apes share a primitive type of episodic memory that is entirely different from the reflective processing ability that occurs instinctively in human adults. As Suddendorf and Busby (2003) note, humans frequently engage in mental time travel, reliving past events and pondering what the future may hold. Despite much experimentation with other animals, they believe that only humans have episodic memory. Other animals learn from single events, but can not be shown to indulge in the reconstruction of past events in order to influence current or future behaviour. Suddendorf and Busby point to observations and experiments with food caching birds. Recovery of cached items appear to only make sense in the light of the bird remembering the original caching event, but they believe that these birds may know where food is hidden without any recollection of any past episodes of caching. Many examples of animal behaviour, like hibernation, nest building, and food caching, appear to involve planning for future events, but Suddendorf and Busby caution the attribution of mental time travel to animals involved in this behaviour. After all, bears hibernate without having yet experienced winter. *H. erectus* did not modify their environments as far as the building of shelters or fireplaces (Noble and Davidson, 1997). Johnson (1989) believes that our hominid ancestors did not organize the world in the same way as *H. sapiens*. Their stone tools were for immediate use and their pattern of food consumption indicates little future-oriented behaviour. It seems that mental time travel, and the episodic memory that this entails, was a recent and unique emergence in hominid evolution.

7.3   Precursors of Communicative Systems

Marcus (2006) is cautious about treating language as a set of modules that have evolved "*sui generis*". Rather, he suggests that we take Darwin's idea of descent with modification into account, when studying the evolution of language, and look to pre-existing machinery in other animals to find where modifications may have arisen. Fitch and Hauser (2004) also believe that "the evolution of the language presumably involved the incorporation of some ancestral primate cognitive abilities".

There is no doubt that some of the physiological systems supporting speech have evolved in other animals. For example, Fodor has pointed to modular systems supporting phonetic discrimination. Animals appear to be able to discriminate speech and non-speech sounds, so this previously evolved system may have led to a distinct psychological faculty in humans that equates with a computational neural architecture for automatically analyzing speech sounds (Fodor, 1983).

Several years ago Hauser et al. presented a controversial paper that explored what we might consider to be the unique qualities of the language faculty. They suggested that the human language faculty might contain just one novel emergent characteristic (recursion), and this Faculty of Language in the narrow sense (FLN), may have evolved *for* other reasons than communication, such as computational systems for number, navigation or social relations. The faculty of language in the broad sense (FLB), on the other hand, encompasses other systems that appear to be part of, or at least support, the language faculty. The FLB they surmise, would include the sensory-motor system, the conceptual-intentional system, and memory, along with less obvious, but necessary, systems of respiration, digestion, and circulation. Implicit in this paper is the assumption that communicative systems already operate in some of our primate relations, even if they appear to be of a different nature to human language. They favour the comparative approach, believing that many of the traits associated with FLB may have evolved as modular systems in other animals, but have become domain-general in humans, perhaps "guided by particular pressures, unique to our evolutionary past, or as a consequence (by-product) of other kinds of neural reorganization" (Hauser et al., 2002).

In a follow up paper, Hauser et al. (2007) revisit the topic of 'Evolutionary linguistics' (*evolingo*), reinforcing the idea that the way forward may not lay with precursors of actual communication systems in other animals. Instead, we should look to the "underlying computational mechanisms subserving the language faculty and the ability of nonhuman animals to acquire these in some form". They are enthusiastic about the prospect of finding precursors of semantics in "non-linguistic conceptual representations", like single-plural distinctions. In addition, they focus on statistical and rule based problems that other animals appear to follow and try to link these to possible precursors of syntax. It is mooted that animals may deploy computations in other realms, like spatial navigation and social interaction, and although these computations may play no part in communicative acts, they may have been co-opted by humans for language. This is a good point. Wallace (1989) has proposed that the sub-cortical regions of the mammalian brain responsible for maintaining cognitive maps of their worlds may have been co-opted for enhanced human spatial-mapping abilities and also for language.

According to Hauser et al., *evolingo* should focus on all known animal computation systems and move away from the "taxonomic myopia" of focusing purely on non-human primates. Confining ourselves to language experiments with just non-human primates has resulted in so many disappointing outcomes, a point these authors recognize. Although they accept the importance of descent from common ancestors, they would like to see more work on distantly related species, which "might share common ecological or social problems, thereby generating common selection pressures" (Hauser et al., 2007). We run into trouble with this approach when we consider that our ho-

minid ancestors arose in *exactly* the same ecological situations
as non-human primates and presumably faced *exactly* the same
social pressures. If selection pressures were responsible for the
emergence of some of the precursors of language, why would
we expect them to not operate on chimpanzees, orangutans, or
gorillas? Pruetz and Bertolani (2007) have studied chimpanzees
living in similar environments to those of early hominids. These
diverse habitats include woodland, grassland and patches of
forest, and although they reveal interesting facts about chimpanzee social and hunting behaviour, there are no apparent
insights into what 'selection pressures' may have led to a change
in cognition or a 'need' for enhanced communication.

As argued previously in this chapter, it is a most difficult task
for those dealing with the evolution of language to find any
precursors in the animal world for any of the supporting functions of language. Chomsky (2005) also believes that there is
no obvious rationale, and no evidence for precursors or earlier
stages of language evolution or its acquisition. Accordingly,
Chomsky (1988) is critical of many ape studies, as he believes
that language is based on entirely different principles from animal communication systems. He is not surprised that language
experiments with apes are a "failed experiment". Most of us
take it for granted that language confers an enormous advantage
and he rightly notes that it is "hardly likely" that chimpanzees
just "never thought to use it" if they are endowed with this
ability. Animal communication, according to Chomsky (2005)
is a representational system based on a one to one relation
between the mind/brain processes and an aspect of its environment, whereas human language is both discrete and infinite. He
points out that even the simplest words and concepts of human
languages do not need this relationship to a world outside of
the mind.

Jenkins (2000) is skeptical of the whole approach toward forming theories of the evolution of language as if there was a trend
toward complexity, 'toward' human language. He believes that
we should focus separately on specific systems like trail marking,

animal vocalizations used in the wild, or any other communicative trait of animals, from studies of human language.

Nevertheless, the quest to find the precursors for human language in our primate 'cousins' pervades the literature in the cognitive sciences. The following sections will outline many of the theories that claim to be the possible starting point for language in our hominid ancestors including animal calls and protolanguage, both vocal and gestural.

7.4   Animal Calls

Along with the assumption of the gradual evolution of language
for communication, there is an underlying notion that human
language evolved from animal calls, which are often considered
to be communicative. In this section, I challenge the idea
that 1) animal calls are communicative and 2) animal calls are
precursors to human language.

Bickerton (2002) asserts that the idea that language evolved
seamlessly out of animal communication systems is an unwarranted assumption that all complex systems must have evolved
gradually and incrementally out of simpler systems. Although
chimpanzees are considered to be our closest living genetic
relatives, our communication systems are dramatically and obviously different. In fact, many other classes of animals appear
to exhibit a more complex form of communication than our
primate relatives. Bees, for example, have evolved not only a
complex social system, but also a seemingly complex communicative system, although Chomsky (1988) warns that so-called
bee language is a misleading metaphor, as bee motion is not
discreet, which is a key prerequisite for language.

Nevertheless, many scholars are convinced that the starting
point for human language may lay in animal calls. Cheney
and Seyfarth (2005) believe that nonhuman primates may be
able to extrapolate complex information from their social group
through their vocalizations, which they believe are discreet and
hierarchically structured. However, Fitch and Hauser (2004)
have found that monkeys have a fundamental computational
limitation in their ability to hierarchically organize acoustic
signals, which is a crucial requirement for the acquisition of any
human language. On the positive side, Rauschecker (2005) has
used MRI techniques to show that there is an homologous area in
the brain (anterior superior temporal) of humans and monkeys
that becomes active in monkeys when they recognise species-
specific calls and in humans when decoding speech sounds.

However, as argued later in this section, this brain activity may simply be the common ability of most animals (including humans) to be able to discriminate the vocalizations of various members of their social group, due to the minor differences in the anatomy of the vocal tract of individuals.

Some scholars consider some of our more 'primitive' utterances to be similar in function to other primate calls. For example, Jackendoff (1999) considers that basic exclamations like *ouch* and *wow* together with utterances like *shs* and *psst* could be likened to primate calls.

Bridgeman (2005) likens primate alarm calls to the first utterances of human infants. He believes that primate calls can be paraphrased into 'holophrasic' cries like "I'm angry" or "Leopard", which he believes are intensional states *for* communication. However, evidence is very thin on the ground supporting the view that other primates have a rich theory of mind entailing the sharing of common intensional states. Povinelli and Barth (2005) have found that what we assume is complex information, requiring a highly complex mind in nonhuman primates, is simply behaviour resulting from behavioural cues. These automated reactions have been instilled in primates after many bouts of experience. As noted by Povinelli (section 7.2.3), human minds can't help recreating the chimpanzee mind in their own image. Pika et al. (2005) have also shown that most communication among chimpanzees is just ontogenetic ritualization. Their inventory of behaviours follow instinctual routines restricted to fixed repertoires, and vary little according to context.

Nevertheless, Bridgeman (2005) believes that from these presumed 'holophrasic' primate calls, a large lexicon would develop, eventually leading to the need for the "power" of modern language. Also taking the gradual approach, Studdert-Kennedy (2000) thinks that the properties of reliable transmission from one generation to the next have enabled language to evolve "in perhaps no more than some tens of thousands of generations, from inarticulate cry to articulate speech".

Bridgeman paints a cosy picture of "some group of *Homo erectus* sitting around their fire after a hard day of hunting and gathering" when someone announces "Lake cold" and another replies "Fishing good". The problem with this picture is that it has no basis in reality. As pointed out in section 5.5, our *Homo erectus* ancestors did not appear to control fire or have fireplaces to sit around for a chat after a hard day's hunting and gathering. They appear to have not even indulged in any hunting and foraging activity that would require any more complex planning than that which we find in extant chimpanzees. The fishing may have been good, but the archaeological record suggests that the ability to fish (using tools or structures like fish traps designed for the purpose) is a unique ability of *Homo sapiens*.

The major question to be settled is that of deciding whether non-human primates engage in intentional communication. Many studies of chimpanzees in the wild reveal a lack of complex vocalizations in chimpanzees and gorillas (Snowdon, 1993). This is despite the fact that their vocal tracts are capable of producing quite a few of the phonemes of human languages. Apes simply do not use any calls, grunts or gestures that could be interpreted as sharing any communicative behaviour. Their calls reflect the knowledge or emotion that the caller has, and *not* any knowledge that they intend the listener to acquire (Cheney and Seyfarth, 2005).

There may be a simple explanation for the assumption that other animals can extract information from the vocalizations of their social group. Owren and Rendall (2001) note that most animals exhibit subtle anatomical variation in both size and shape of the supralaryngeal cavities, and this allows an individual to produce vocalizations that mark its identity. Humans have retained this ability to discriminate the tonal variations in human speech and immediately identify the speaker from many thousands of other people.

Much work has been undertaken in order to discover homologous areas in the human and non-human brain, that might be involved in vocal recognition. There appears to be a homologous area in the brain where monkeys recognise species-specific calls, and where humans decode speech sounds (Rauschecker, 2005), however, Barrett et al. (2005) find that most of the anatomical regions that control language functions in the human brain do not have homologues in other primates. They have found two distinct functional regions within Broca's area, both in gross morphology and cytoarchitecture, one controlling the higher language functions like syntax and lexical semantics, and the other articulatory and motor speech functions. Anwander et al. (2007), using magnetic resonance (MR) imaging, have detected areas not only of high anatomical and functional complexity in Broca's area, but also patterns of long-range connectivity.

Panksepp (1998) makes the point that the vocal signs of social communication in animals are quite different from human propositional speech. Vocalizations in monkeys are typically organized by sub-cortical brain circuits which relate to the generators of emotional states. Animals with little neocortex, he argues, are driven by their emotions, which are embedded in the ancient organization of the sub-cortical regions known as the extended limbic system.

Supporting Panksepp's findings, Ploog (2002) has also identified the neural basis of speech vocalisation, which is quite different in humans from where other primates generate calls. He has identified the newly evolved neocortical connections that are involved in the production of human language. These emergent pathways are not present in other primates, however humans retain the limbic emotional input necessary for vocal expression, and damage to either of these systems results in impairment to speech. Limbic structures are always involved in speech and can usually be identified by the expression of emotion that accompanies our words. The human brain is quite different from other animals with massive interconnectivity, where every part can find an access pathway to any other part. The right hemisphere in humans has a cortical zone corresponding to Wernicke's area, which has been shown to be involved in the interpretation of emotional "tone" of language (Ploog, 2002).

Fitch (2005) points out that the key difference between human and ape vocalization may be the long-distance corticomotor connections from the premotor cortex to auditory motor neurons. These newly evolved connections, he suggests, appear to be the vital link that gave hominids voluntary control over vocalization.

In summary, we find that animal calls are indexical and not symbolic. Animal calls are meaningless or misleading in the absence of an immediate referent (Bickerton, 2007). In contrast, symbols can refer to anything in a human's experience, whether present or not, or even if it refers to something that is non-existent in the 'real' world. Due to the fact that we cannot find a plausible starting point for symbolic communication in other primates, and there is scarce (if any) evidence for symbolic behaviour in any of our hominid ancestors, it becomes increasingly difficult to argue for a gradual emergence of a complex communicative system during our long evolutionary history. Throughout the following sections, I will reference much of the research that points to the key differences between human brain architecture relating to vocalization and speech recognition, and that involved in animal vocalization.

Vocal learning and Primate Vocalization

The ability for vocal learning[8], which is essential for the acquisition of human language, can be found in three distantly related groups of mammals (humans, bats and cetaceans) (Jarvis, 2004). This ability can also be found in three, also distantly related, groups of birds (songbirds, parrots and hummingbirds). Jarvis believes that there must be something special about the brains

---

[8]Piattelli-Palmarini (1989) prefers the more plausible term "parameter setting", to describe how both humans and other animals 'learn', due to fact that the term 'learning' carries strong implications that some sort of 'instruction' is going on.

of vocal learners, as all vocal learners (e.g., humans, songbirds) have very close relatives who do not show any aptitude for this instinctive trait (e.g., apes, pigeons).

Jarvis finds that vocal learning is quite different from auditory learning. He uses the example of teaching a dog to understand the word 'sit'. A dog learns to 'hear' and react to the word 'sit', but cannot imitate the word. Likewise, Klinowska (1994) is unimpressed with studies of cetaceans (whales, dolphins and porpoises). She thinks that any performances are better described as just ordering responses for some reward, and are not related to any rudiments of language. Their echo-location systems are an aid used only for hunting and exploring their environments, but she believes that their repertoire is far too limited for the understanding and use of language.

Among primates, only humans have the neural pathways for vocal learning. Anderson and Lightfoot have pointed out that

> non-human primates certainly have ears to hear with
> and (with more qualification) a vocal tract capable
> of producing sound, or hands and eyes capable of
> producing and identifying signs, but this does not
> appear to endow even the cleverest of them with
> the capacity for human language (Anderson and
> Lightfoot, 2002).

Savage-Rumbaugh (1994) agrees that apes have the necessary auditory system to recognise human speech sounds, but they cannot produce them. Also critical to understanding human speech is the ability to recognise consonants, as they provide the categorical boundaries that segment speech sounds into distinct and meaningful units. Apes cannot produce consonants, so their communicative vocalizations are composed entirely of vowel sounds (Pagel, 2000). Savage-Rumbaugh is more enthusiastic about the abilities of chimpanzees. Her work with the bonobo Kanzi has led to the belief that he is able to decode consonants in human speech, even if he cannot produce them. In fact she goes as far as saying that "if he could produce consonants as well as vowels, and if he had the requisite degree of neurological control, I have little doubt that Kanzi would be able to speak" (Savage-Rumbaugh, 1994). Could we not also suggest that if Kanzi had wings and a syrynx, there would be little doubt that he could fly and sing like a bird?

Piattelli-Palmarini (1989) has found that a set of parameters operates for the phonological aspects of language. These parameters appear to be similar to Chomsky's (2005) "switch box" for the *Principles and Parameters* model of syntax. The physiological articulators of the vocal tract specify a limited set of admissible movements that belong to the particular language to which a human infant is exposed. Once the phonological parameters are 'set', they are rarely violated. Piattelli-Palmarini points out that although children sometimes invent new words, they do not violate the phonological boundaries of their native language. In addition, children do not 'take on' the accents of the second language of their migrant parents. Each E-language has its own specific phonological boundaries. Rather than having been honed to accommodate speech, he finds that vocal tract physiology provides a mechanism of 'selection' for a limited set of phonological parameters.

Further evidence for an innate 'selective' mechanism comes from the study of phonetics. Jenkins (2000) puts forward the interesting example of the distinction between the *r* and *l* sounds, which are included in the available universal phonetics for all languages. Japanese infants naturally recognize this distinction, but due to the fact that the Japanese language system does not utilize the distinction between these two sounds, this ability to discern a difference atrophies during the course of their language development.

All infants "babble" regardless of the language/s to which they are exposed[9]. Petitto and Marentette (1991) have shown that,

---

[9]Manual babbling occurs in deaf children exposed to signed language

due to the similarities between manual and vocal babbling, we cannot assume that infant vocal babbling has anything to do with the maturation of the articulatory apparatus. Rather, babbling occurs as an expressive capacity that is brain-based, and is linked to an abstract linguistic structure, which is capable of processing different types of signals in the signed or spoken modality. We can see that this capability in human infants is a 'far cry' from animal vocalizations.

Vervet alarm calls

Vervet calls are often used as an example that some primates, other than humans, have evolved a communication system. For example Dennett (1991) argues that vervets evolved a 'necessity' of communication in order to deal with the stressful lives they lead. However, Owren and Rendall (2001) warn of the dangers of analyzing vervet alarm calls in terms of human linguistics. They argue that the acoustic structure of these calls tap into affective areas of the subcortical regions of the primate brain and perform simple nonlinguistic functions. Rather than having semantic content that a vervet wants to share with its conspecifics, the alarm calls serve only to 'release' behaviour in others. They observe that these affective states are similar to the human condition of emotional contagion. Humans sometimes appear to be 'catching' another's affective experience or expression, as in laughing or yawning. We also may execute a conditioned response to seeing a police car or hearing a gun shot. These startle effects in humans access low-level mechanisms of arousal and are produced by the nervous system in the subcortical regions in the brain, mainly the brain stem. Vervets produce vocalizations that contain the same acoustic features and recruit these same brain mechanisms for 'startle' reactions. Most calls consist of a rapid series of abrupt-onset, broadband pulses with high overall amplitudes.

---

from birth

Owren and Rendall (2001) note that human babies along with other young animals unknowingly exploit this same mechanism to influence the attention of carers. These calls do not have rule-based sequencing and do not contain compound signals. Rather, as a result of Pavlovian conditioning, they induce emotional responses in most animals, including humans. They suggest that the intuition that vervets use a 'sign' to share information among the group most likely stems from our mistake of anthropomorphizing primate behaviour.

Knight (2000) supports this stance noting that primate vocalisations are irrepressible and context-bound. Human speech functions as a communication device that allows for the sharing of information that is independent of currently perceptible reality. Signals for animal vocalisations link directly to the laryngeal motor neurons. Humans have developed enlarged cervical and lumbar regions allowing for new pathways in our corticospinal region linking to motor neural pools. We devote part of the cortex to representations of the larynx, pharynx, tongue and mouth, necessary for the control of speech. Apes and monkeys do not use these areas for phonation.

Vervet calls are often viewed as reflecting concern for the safety of the other members of the group, but Owren and Rendall have found that they are actually "sender-centered". Often a caller will vocalize when it perceives itself to be in danger, but an animal may remain silent, even when its own kin may be in danger. The point is, vervet calls are not meant to convey information to others, not even between mothers and infants. Animals may appear to be engaged in complex signaling, but are actually just experiencing shared emotional states. Young vervets do not have to learn these calls *de novo*, but capitalize on calling tendencies that are hard-wired into the circuitry of the limbic-system. Limbic fear circuits produce appropriate fear alarm calls that are common to many species and are therefore easily assimilated by young members of the group. Owren and Rendall also point out that many studies on primate acquisition of vocalizing bear no similarities with language acquisition in human infants. When infant vervets first hear an alarm call, they do not execute an escape action, but they learn eventually to respond over time, sometimes in response to the experience of an attack.

Snowdon (1993) suggests that animals in general appear to acquire escape behaviours as learned responses to emotionally driven signals associated with different predators. This does not imply that the animal is actually symbolizing the predator. In addition, vervet monkeys are egocentric. It has been shown that vervets do not understand the minds of their companions, which is similar to the situation with apes.

With auditory events it is extremely difficult to find evidence of vocal learning in any other primate species, even though most primates have quite good abilities for categorization of sounds (Snowdon, 1993). Most, if not all, primate vocalizations are genetically controlled rather than learned. The only evidence of 'learning' that can be observed is in the perception and appropriate use of calls, where novices copy their elders. They become conditioned to respond appropriately to martial eagles, to snakes and to leopards.

Baboon calls

Baboons generally live in highly organized matrilineal social groups and when moving through their range appear to give loud *contact* barks. Cheney and Seyfarth (2005) observe that baboons often appear to be answering one another with perhaps the intent of informing one another of their location. This however is not the case. It appears that baboon 'contact' barks may simply reflect the signaler's state of separation from its group. Callers vocalize when they are anxious, and listeners can extract the location of the caller from this information. This implies no intention on behalf of the caller. Experiments with loud speakers emitting fake distress calls show that a baboon will only vocalize when they are themselves separated from the group, and not when they hear a call from a relative. Calling is simply an emotional response to an emotionally stressful situation. They conclude that this difference in communication behaviour underlies a fundamental difference between human language and nonhuman primate cognition.

Horwitz et al. (2005) believe that the difficulty in training monkeys to recognise and categorize auditory sounds is most likely due to the difference in neurobiological architecture. They point out that humans can discriminate a huge number of environmental sounds in the order of 10^5 and this extension of auditory neurobiology may be the key to understanding human language evolution.

7.5   From Animal Calls to Protolanguage

For many scholars, the starting point of language must have been the emergence of a protolanguage, either spoken or gestural. This assumption is often made with an appeal to 'common sense' (Macneilage and Davis, 2000), or that language must have evolved with a first 'step' (Fitch, 2005). If language evolution was gradual, then it must have had a starting point that was much simpler than a full language faculty. An additional temptation, which according to Bickerton (2005c) is unwarranted, is to equate a protolanguage for early hominids with the simple words that human infants acquire. Human infants are often deemed to recapitulate the stages of our hominid ancestors during language acquisition. I explore this approach further in section 7.6.

Although Bickerton rejects the recapitulatory approach for the evolution of language and maintains that fully syntactic language arose only with *H. sapiens*, he nevertheless believes that the complexity of modern syntax *must* have had the precursor of a simple protolanguage, which may have involved only single symbols or words. He believes that "simple logic dictates that symbolic units must exist before any way to concatenate these can come into existence" (Bickerton, 2007).

However, Bickerton's claim that protolanguage arose to convey symbolic concepts leads to a contradiction. Elsewhere, Bickerton (2005a) has pointed out how ambiguous single utterances can be. He uses the example of someone shouting "fire", and how difficult it would be to decipher the intentions underlying this utterance. With this example, he claims that to extract any meaning would be to first have acquired a rich conceptual system of fire in all of its contexts. He accuses others of "smuggling" in a "ready made" world of language-like ability from a presumably holistic utterance, but it is difficult to pin down just what *his* protolanguage would entail.

A "scavenging niche" of hominids around 2-3 millions years ago is claimed by Bickerton (2007) to have been the likely selective

pressure that "obligated" them to return to base camp and
inform their fellow camp-dwellers of their finds. Single-unit
utterances (words) represented concepts, and their transmission,
he believes, would have paid off in terms of survival. Strangely,
he states that the reason a proto-language has not arisen in any
other seemingly intelligent creature is that there was "no compelling pressure selected for such a use of conceptual structure
by members of those species"[10].

Although Fitch (2005) does not want to appeal to "common
sense" for earlier speech stages, he nevertheless argues that
some type of "protolanguage" is a necessary stage in language
evolution. He thinks that the evolution of long-distance corticomotor connections from the premotor cortex to auditory
motor neurons may have been the "crucial neural step" that
gave hominids the voluntary control over vocalization.

7.5.1   Imitation

It is often thought that the ability to imitate may have been a
crucial breakthrough in hominid cognitive and language evolution. Richerson and Boyd (2000) believe that human cognition
has mainly evolved to acquire and manage cultural traditions
that are made possible by a major adaptive breakthrough allowing for the capacity for imitation. A theory of mind is thought
to have come into being for social purposes, and imitation is
thought to be part of this capacity. For Arbib (2005), mirror
neurons, together with orofacial gestures, are deemed important
for the ability to imitate and may have provided the scaffolding
for the emergence of proto-language. Mirror neurons were first
discovered by Rizzolatti et al. (1996). They found that an
area of the monkey premotor cortex [F5] contains neurons that

---

[10]Previously, as cited in this chapter and elsewhere in this work, Bickerton
(1998, 2002, 2005c) has been very scathing of 'special case' pleading for
the evolution of language in hominids. It therefore seems strange that he
believes that a protolanguage arose in hominids without any modification
of the ancestral primate brain (Bickerton, 2007).

appear to be active both when the monkey actually performs an action *and* when the monkey sees another monkey perform a similar action. Activity was also found in the [F5] area when the monkey observed the experimenter performing a similar action. It was proposed that the monkey was 'representing' the observed action. Because humans appear to have an anatomical homologue in the Broca's area, which also becomes active when observing a grasping action, they suggest that the functional specialization of Broca's area derives from an ancient mechanism involved in the production and understanding of motor acts. They are sympathetic to the idea that verbal communication systems may have gradually evolved from the recognition of hand and face gestures.

Tomasello et al. (2005) have performed some interesting experiments to test whether chimpanzees may achieve more success in solving a problem involving imitation when they have an understanding of the causal relations in an apparatus. They conclude that even when causal mechanisms seem to be reasonably opaque, apes still do not imitate, but engage merely in emulation learning. Emulation of a task can result in a similar outcome, but involves little or no understanding of the method of achieving the result.

Chimpanzees, according to Tomasello et al., have many opportunities to copy from others in their own group as well as their mothers, but they do not appear to have this ability to learn through imitation, which requires some sort of causal understanding of the task at hand. The main problem with learning through emulation is due to the fact that the task does not allow for extension into other arenas. It does not allow for creativity or flexibility, which generally means that the task just becomes repetitive ritualization.

Although imitative behaviour leading to a powerful form of cultural learning naturally emerges in human infants by around 15 months of age, an inherent problem comes with the claim that the ability for imitation was an important and necessary precursor for the acquisition of language. Piattelli-Palmarini (1989) claims that the ability to imitate has very little to do with language acquisition. Rather than imitate their peers or parents, children come to language equipped with innate selective mechanisms, which have little to do with learning by instruction, or with the ability to imitate. As noted elsewhere, children do not 'imitate' the accents of their migrant parents, and they do not adjust their grammar according to a parent's correction – quite simply, they do not 'learn' their language from their parents, nor do they 'imitate' the language of their parents (Piattelli-Palmarini, 1989).

7.5.2   Gestural Imitation

Due to the fact that earliest hominids did not appear to have the necessary anatomical foundation to support spoken language, many scholars have asserted that a proto-language emerged initially as a gestural system. There have been many arguments put forward claiming to have ascertained the 'reason' for a protolanguage.

Armstrong et al. (1994) claim an origin for gestural language going right back to the first hominids. For them, "it is clear beyond any reasonable doubt that it was the shift to bipedalism that defined the origin of the hominid lineage", and it was this modification to upright posture that allowed for the freeing of hands for gesturing. Due to the fact that much of sign language employs gestural icons that appear to contain an inherent syntax, they suggest that these may be the building blocks of syntactic language. Chimpanzees, according to Armstrong et al., can be trained to communicate objects and events through the use of symbols and icons, and they argue for a continuity of this ability between chimpanzees and the first hominids.

However, Pagel (2000) observes that "only after years of almost continual Skinnerian training (harassment?) do chimpanzees show a limited facility with sign language word use, and even then, little or no concept of grammar". Moreover, as we learned earlier, symbols used by chimpanzees are meaningless or misleading in the absence of an immediate referent (Bickerton, 2007).

Bodily Mimesis

Zlatev et al. (2005) have put forward a theory of bodily mimesis as the key to understanding what may have been the necessary precursor for language. A gesture, under conscious control, can stand for an action, an event, or an object. They argue that "bodily mimesis" builds "mimetic schemas", which are prelinguistic but representational and emerge through imitation. Although they do not expand on their statement that this mimetic hierarchy can be applied to ontogeny as well as phylogeny, we can see here their implication that human infants recapitulate the 'stages' of evolution of ancestral hominids. Their model sees this ability emerging gradually in four separate stages.

The first stage they call "protomimesis" involving "body matching" that underlies the neonatal mirroring in which both human infants and chimpanzees engage. This first stage, they believe, lacks volitional or conscious control. The second stage is "dyadic mimesis", which involves conscious control over the body's representational movements. This capacity allows for "mirror-self recognition" – a skill underlying the ability to understand that one entity can correspond to another. By 14 months, children pass the mirror self-recognition test and can perform mimes of previously observed events. Although apes have some difficulty with these tasks, they have some ability for "dyadic mimesis" (Zlatev et al., 2005).

The third step in the "mimetic hierarchy" is what Zlatev et al. believe to be crucial for the grounding of a language capability. This step involves recruiting intentional body movements for communication with others. Pointing, and understanding what this means, would be an example of this stage. Due to the fact that these mimetic schemas can be "interiorized", this ability may be the grounding for a child's first words, in particular, verbs. Because these "signs" are derived from imitating culturally relevant actions and objects, they can be shared intersubjectively. The fourth and final step involves fully conventionalized and systematic signs that may be characterized by languages such as American Sign Language.

The problem remains. As Aronoff et al. (2008) point out, there are no documented cases of chimpanzees, bonobos, orangutans, or any other primates in the wild, engaging in the use of the body or the hands to represent anything. Although Zlatev et al. believe that chimpanzees can be trained to understand that one entity can correspond to another, this is only after many bouts of training (see Pagel (2000)). Hominids did not have the luxury of this training, so it is difficult to support the idea that the crucial step to "mimetic hierarchy", involving intentional body movements for communication, would have arisen spontaneously.

Probably the most controversial theory comes from Arbib (2005) who argues that a full-blown syntax and compositional semantics were a historical phenomenon within the development of *Homo sapiens*. Arbib argues that fully formed languages, as we know them today, involved few, if any, further biological changes from our hominid ancestor. He believes that the "initial seed" for a communicative system may lie in manual gesture involved in pantomime, which was made possible by primate mirror neurons. Mirror neurons appear to be active for both execution, and observation of manual actions. Arbib suggests that other communication "devices" like primate involuntary calls and facial gestures registering emotional states, may be seen as signalling certain aspects of current internal states. Mirror neurons, together with orofacial gestures, are deemed important for the ability to imitate.

Because both monkeys and humans contain an homologous

"mirror system" (in the premotor area F5 in monkeys and Broca's region in humans), we may find in this system the "scaffolding for the emergence of protospeech" (Arbib, 2005). Brain imaging studies show that F5 and the homologous region in Broca's in humans are activated when both species execute a grasp, or observe another executing a grasp. Rather than speech systems evolving first, Arbib argues that manual gesture, afforded by the mirror system, allowed for complex imitation of sequences, which is possible after training in chimpanzees, but comes naturally to humans. It seems that mirror neurons may play some role in human cognition and language. Studies show that humans with autism show decreased connectivity in the frontal area 44, which according to Wallace (2005) is compatible with the hypothesis of mirror neuron defects associated with autism.

Arbib (2005) believes that a simple pantomime could have gradually evolved as a "expanding spiral" into a complex "open repertoire of manual gestures". When *Homo sapiens* emerged, they apparently had evolved the "neural critical mass", which meant that they were biologically "language ready" for full-blown syntactic speech, but acquired it through "cultural innovations". Protosigns, he argues, became ever more holistic and flexible covering a wide variety of experiences. The control systems that allowed for ever complex protosign are believed to have eventually come to also control, with increasing flexibility, the vocal apparatus.

Corballis (2002) also believes that gestural language was the precursor for vocal language, with the switch from gestural to spoken language taking place earlier than Arbib suggests. Corballis thinks that the switch would have taken place, presumably with the emergence of *H. habilis* or *H. erectus*, and may have been "necessary" in order to free the hands for making stone tools.

A sympathetic view of the theory of gestural origins of language is taken by Emmorey (2005), but she cautions against the proposal that spoken language evolved from protosign. Some modalities in sign language are derived from nonlinguistic gestures such as clenching the fist and deictic pointing, but this does not mean that this sort of communicative gesture is a remnant from the past. It should be made clear that we are dealing with modern sign languages acquired and created by modern human brains. She reminds us that signed languages are just as complex and efficient as spoken languages, and we need to explain why spoken language appears to take precedence in evolutionary theory about language origins, when the potential of choking to death exists as a result of the repositioning of the larynx[11]. Emmorey points out that signed languages are easily acquired by children and are processed by the same neural systems within the left hemisphere.

In support of Arbib's theory, Wray (2005) believes that even much of modern linguistic communication can be considered holistic. Using evidence from semi-literates, Wray suggests that we take in language using a needs-only analysis, and in a holistic manner. The oddities and arbitrariness of language can be explained by this approach, as language has evolved purely to meet expressive need and not as a strictly ordered or systematic entity. It is also suggested that the inherent regularity of pattern that linguistics finds may be put down to culture-centricity and the emergence of formulaic speech in the absence of literacy and formal education. In other words, most people, especially in pre-literate days, which represents most of human evolution, have produced and perceived language in a holistic manner, and not in a strictly analytic way. For Wray, early 'language' would have involved fixed phrases used to command, request, or warn. Single words would represent single concepts with all its 'grammar' embedded within it. Later in this section, I offer arguments that undermine Wray's theory. Wray is accused of 'smuggling in' a 'ready-made' world of concepts, or even

---

[11]Although, to my mind this danger is overstated as I'm sure there were many more life-threatening hazards for hominids than choking on their food. A few choking casualties would not have had a large impact on the survival of the species.

grammar, into her 'holistic' language era (see Bickerton, 2005a; Fabrega, 2005).

The "Mirror Neuron" theory for protosign or pantomime has been severely criticized. For example, Barrett et al. find behavioural and anatomic studies inconsistent with this theory. They find little anatomic evidence supporting a link between Broca's area as a shared neuronal substrate for human gesture and language. In addition, they challenge Arbib's idea that hominid pantomime may have also co-opted primate orofacial gestures and primate vocalizations to communicate emotional state. While the left hemisphere in humans usually controls propositional speech, it is the right hemisphere that controls vocalization associated with emotional prosody. Hence, "this double dissociation argues against left-hemisphere dominance for comprehending, imitating, or producing emotional facial expression or prosody" (Barrett et al., 2005).

Further criticism comes from Fabrega (2005), who is mostly critical of Arbib's avoidance of the issues dealing with the evolution of many of the cognitive abilities that would have had to underpin even the simplest form of language, like mind, consciousness, and self. For Fabrega, any theorizing about protosign or protospeech would have to also involve a theory of protoculture. He adds that theories of protosign or protospeech already presuppose enormous cognitive capacities. Capacities like conscious awareness of self and the situation, together with the ability to decompose and order a complex perceptual scene, would have to already be in place before the emergence of even the simplest holophrasic utterance or gesture. Arbib imagines a spiraling effect of ever more complex sign or speech eventually leading to full language ability in *H. sapiens*, but Fabrega asks "how much of the protosign/protospeech spiral is enough?"

> *If* there is a shared body of knowledge about what pantomimes are for and what they mean, what disambiguating gestures are for and mean, and what speech sounds are for and mean, *then* there ex-

ists an obvious meaning-filled thought-world or context "carried in the mind" that encompasses self-
awareness, other-awareness, need for cooperation,
capacity for perspective-taking - and, presumably,
a shared framework of what existence, subsistence,
mating, parenting, helping, competing, and the like
entail and what they mean (Fabrega, 2005).

Tomasello et al. make the important point that understanding
intentions may be the most important step for the evolution
of human cognition, and importantly, for language acquisition.
Understanding intentions of others is foundational for deciding
precisely what the other is aiming to show. For example, the
same physical movement may be seen as "giving an object,
sharing it, loaning it, moving it, getting rid of it, returning
it, trading it, selling it, and on and on" (Tomasello et al.,
2005).

McNeill et al. (2005) are dubious about the whole gesture-
first paradigm of language evolution. They feel that theories
supporting a gesture first origin of language wrongly presuppose
that gestures are the simplest form of language, and also that as
humans evolved, their language systems became more complex
until speech supplanted gesture. Rather than this replacement
hypothesis, McNeill et al. believe that gesture and speech
most likely co-evolved as a thought-hand-language circuit in
the brain, which appears to be the system in modern humans
today when they combine these two elements to jointly convey
their message.

A further criticism of the gestural first theory comes from
Bickerton (2005a), who doubts very much whether hominids
had the cognitive powers necessary to engage in complex mime
and holophrasic protolanguage. He points out that even today,
playing charades is quite difficult for humans. Even with the
resources of modern human language and cognition, we need to
resort to stereotypic gestures for "film title" and "book title",
and counting on our fingers for the number of syllables or words.

We often find it impossible to guess what the pantomime is trying to portray. He questions whether any animal could extract a "situation" from the unbroken stream of pantomime experience without a language. Arbib is accused of introducing in a ready-made world of language where his holistic protosign stage seems unnecessary. For example, Bickerton points out that to extract a symbol meaning "fire" from a holistic utterance, we first need the semantic concept of fire in all of its situations. The same would be true of an holistic gesture. The dismissal of syntax as having arisen historically, after *H. sapiens* emerged, as posited by Arbib, makes the argument implausible for Bickerton. He believes that any theory that fails to deal with the two substantive issues in language evolution, how both symbolism and syntax emerged, should not be taken seriously.

Rauschecker (2005) questions the importance of considering Broca's area as a stand-alone functional region of the brain. He believes we must consider the complete circuit, which (using MRI techniques) shows that many more brain areas are implicated during the perception and production of language apart from the traditional Broca's and Wernicke's areas. There is also an homologous area in the brain (anterior superior temporal) of humans and monkeys and this area becomes active in monkeys when they recognise species-specific calls and in humans when decoding speech sounds. For Rauschecker, this could equally be considered a starting point for proto-language, instead of Arbib's gestural system based on homologous mirror neurons.

Indurkhya (2005) is also cautious about using mirror neuron experiments in monkeys to extrapolate to humans. For Indurkhya, the monkey experiments show that it is the effect on the object that is important to the monkey, not the physical appearance of the effector. This has implications for the ability of any animal to imitate and project itself into another mind, rather than just emulate or produce behaviour based on imprinting or stimulus enhancement. The ability to project one's thoughts and feelings into another person is for Indurkhya, the crux of the distinction

between humans and other animals. As argued in section 7.2.3,
it is unlikely that early hominids had this ability to interpret a
wide variety of actions and situations. If mirror neurons do not
register the actions of the effector, as Indurkhya claims, then
it difficult to see how Arbib's hominid pantomimes could have
been enacted. Only with the arrival of modern humans do we
find evidence for ritual behaviour indicative of the capacity for
empathy (as discussed in section 5.4).

Probably the most important question to ask is: why, if our
hominid ancestors had evolved an innately embedded gestural
language over many millions of years, does this 'natural' ability
not appear spontaneously in all humans today? Humans use
basic gestures, similar to other primates, but these are *not* dealt
with in the language areas of the brain, just as screams, sighs
or gasps are not generated within the language areas of the
brain (Carter, 1998).

7.6   Recapitulation - Acquisition and Evolution

Commitment to a gradualist approach to the evolution of language appears to underpin the claim that human children more
or less recapitulate the evolutionary steps of our hominid ancestors as their "language" grew in complexity after the split from
the other apes. Speech acquisition in human infants is used to
infer the initial patterns of speech in hominids, and thereafter,
a subsequent increases in complexity.

It is argued by Bickerton (2005c), that there are no grounds for
assuming that the ontogenetic order of language development
mirrors that of its ancestral emergence in hominids. However,
due to the fact that human infants seem to acquire language in
'stages', leading from very simple words to complex expressions,
for some the temptation to equate this sequence with the various
stages of language evolution in our hominid ancestors is often

too much to resist. For example, Bridgeman (2005) believes that language evolved in small steps suggesting that "we can look to human development for hints about how the evolution of language may have proceeded". He believes that a child's first words are holophrasic and are like primate alarm calls.

Macneilage and Davis (2000) also claim that "it is common sense that speech must have been simpler in earlier times than it is now", and use this assumption to support their claim that "the first truly speechlike behavior of infants closely resembles the proposed initial frame stage of the evolution of speech". Their approach typifies the Darwinian incremental view of how complexity arises, gradually and through generational change. For Macneilage and Davis, "successive generations" of hominids were forced to assimilate more complex inventories of speech patterns due to the increased demands for complex communication among adults. They equate the initial frame stage in infants, which involves basic motor capabilities, with the very beginnings of hominid speech capacities. These motor capabilities include mandibular oscillation, consonant-vowel co-occurrence constrained by the tongue and other biomechanics of the oral apparatus. It is argued by Macneilage and Davis, that infants acquire more sounds and sound combinations during the course of speech acquisition in the same way that early hominids were slowly building their repertoire under selection pressure to increase the size of their message set.

Following a recapitulatory path, Jackendoff (1999) has put forward a theory proposing that we can observe the likely path of language evolution, with ever growing complexity, within the language acquisition of today's human infants. He makes several basic assumptions when formulating his theory for the evolution of language in humans, which can be easily challenged. His speculations are grounded in the common belief that language evolved incrementally due to an ever increasing advantage for hominids to communicate with each other effectively. He claims to be able to 'decompose' our modern linguistic faculty into separate modules that evolved independently and he claims that we can find 'fossil' traces of some of these independently evolved modules within the path of language acquisition in human infants. Jackendoff also speculates that we should be able to find more 'fossils' of language in pidgin languages, aphasics, home signers and apes. Moreover, these communication systems "fall back" to what colleagues have named "The Basic Variety" of language competency (Jackendoff, 1999).

Jackendoff states that "certain little-remarked aspects of modern language are if anything more primitive than the child's one-word utterances". Exclamations like *ouch* and *wow* are considered fossils of a more primitive language, and utterances like *shs* and *psst* are likened to primate calls. The analogy of infant speech with primate calls presents a major problem. Firstly, pre-linguistic infants do not use the terms *ouch*, *wow*, *shs*, or *psst*. Also, as noted previously, apes do not even come close to using a call like *psst* or *shs* (Tomasello et al., 2005). Apes do not try to direct attention of an object or event to one another (e.g., *psst*, *wow*), nor do they produce calls to direct the behaviour of one another (*shs*). Animal calls merely reflect the knowledge or emotion that the caller has, and *not* any knowledge that they intend the listener to acquire (Cheney and Seyfarth, 2005). In addition, the neural basis of speech vocalization and animal calls is quite different, with primate calls originating from sub-cortical brain circuits related to the generators of emotional states (Panksepp, 1998). Although humans retain the limbic emotional input for vocal expression, they have evolved neocortical connections that are the necessary basis of speech, and these emergent pathways are not present in other primates (Ploog, 2002).

Pidgin languages used by immigrant workers are deemed to further support Jackendoff's theory of primitive 'fossils' of 'basic communication' to be found in modern language use. However, DeGraff (2001) is suspicious of the notion that pidgins or proto-creoles represent a form of language that can serve as 'basic communication'. He argues that I-Pidgins emerge in situations where limited or abrupt language contact occurs, and in some instances of second language acquisition. He does not see the need to invoke special cognitive processes to describe how these languages emerge. Pidgin languages are often viewed as a sort of 'emergency' language in order to expedite communication between two different language groups. DeGraff makes the important point that this process can be seen as quite a complicated and intellectually demanding task. For a speaker to strip away the complicated nuances of his language in order for a 'foreign' speaker to comprehend his communication, is quite a challenge. When formulating a sentence, one speaker will need to have an awareness of his audience's understanding of the speaker's language in order to create a 'middle ground' for both parties to enjoy a successful rapport. Even in the unlikely case where our hominid ancestors had some sort of 'basic communication', it is hardly likely to be of the same complexity as today's pidgin languages.

Peters (1993) contends that our contemporary "infant-directed speech" is a preserved "language fossil" the same as the form of the simplified ancestral "adult-directed speech". She likens the smaller brain of our hominid ancestors with that of human infants and thus surmises that these early ancestors could only manage to produce and understand this primitive 'talk' in line with their reduced cognitive skills.

The recapitulatory approach is at odds with our understanding of developmental processes in the brain of all mammals. As argued in chapter 4, all primates, including our hominid ancestors, had varying life-histories, including brain growth trajectories. We can see from the fossil record that each of our hominid ancestors, from early hominids and the australopithecines, through *H. erectus*, *H. antecessor*, and our Neandertal 'cousins', had widely varying life-histories including the developmental stages of infancy, the juvenile stage, adolescence and adulthood. There is absolutely no evidence that different life-stages were gradually being 'pushed back' in the sequence of development.

The first hominids had brain sizes the same as extant apes, and the brain size of *H. erectus* was no larger than we would expect for a primate of that body size (Eldredge, 1996; Smith and Tompkins, 1995). The endocasts of most of the australopithecine fossils, even the most recent australopithecines, who lived up until around 1.2 million years ago, clearly indicate an apelike cerebral cortex (Falk, 1992). We have no reason to assume that the trajectory of brain growth in these hominids deviated from their pongid ancestor. Panksepp (1998) points out that brain size is more-or-less metabolically attuned to support the functions of body size, and the energy requirements for "invisible mind-like" processes are very low. We therefore have little reason to equate brain size with complexity of cognitive behaviour.

Rather than concentrate on brain size, it seems more profitable to look at developmental programs of our hominid ancestors. The absolute brain size of *H. erectus* and *H. sapiens* was similar at birth, but they did not follow the same developmental pattern (Hublin and Coqueugniot, 2005). *H. erectus*, our immediate ancestor, appears to have followed a brain growth pattern similar to that of a chimpanzee, and earlier hominids. Human infants have a brain size of about 825 grams by the end of the first year, which is roughly the size of a fully developed *H. erectus* brain (Montagu, 1989). A radically different growth pattern is evident in the *H. erectus* brain, where 84% of adult brain size is achieved by one year old, whereas the human infant has only around 50% of adult size at this age. We can therefore presume that brain development in any of our hominid ancestors did not parallel any of the 'stages' in the human infant. We do not 'recapitulate' language evolution during language acquisition.

Since the demise of the theory of strict recapitulation in the late 19C, the term 'recapitulation' has become somewhat of a taboo expression. The theory that evolution proceeds by the addition of new features onto the *Bauplan* of an existing species has been attributed to the German *Naturphilosoph*, Ernst Haeckel[12].

---

[12]Although Richardson and Keuck (2002) have shown that Haeckel did

Haeckel's "ontogenetic law" stated that "ontogeny repeats phylogeny", although, as Richardson and Keuck (2002) point out, Haeckel allowed for the deletion of whole stages of development, as well as the transformation of stages during development. It is the transformation of stages during development, I argue, that has been the main cause of evolutionary novelty in our hominid ancestors, as well as our own emergence.

Raff (1996) believes that there may be just a "shadow of truth" in Haeckelian recapitulation. Humans, for example, have a tail during a certain time of their embryonic state, but this gets resorbed before birth. Most mammals, including humans also exhibit gill slits in the early part of their embryonic form. Although many species share several phenotypic features at some stages during development, they certainly do not go through all of the stages of development of their parent species (particularly not the adult stages), and then add another feature on to the end of the developmental process. Raff is particularly cautious of viewing genetic evolution in a "Haeckelian" fashion, where new sequences are claimed to have been "grafted" onto the ends of old ones. Rather, new gene sequences evolve by substitution, deletion, recombination and duplication, as well as duplication of bases. He makes the important point that evolution erases information as it creates new or changed gene sequences.

There is no doubt that the mammalian brain exhibits some recapitulatory sequence during development. McKinney (1998) points out that humans seem to share many traits with other taxa as the triune brain develops. For example, the first motor displays in human infants resemble the ritualized motor behaviour in reptiles as the basal ganglia matures. In addition, as the limbic system matures, more complex emotional behaviours like bonding and attachment appear in most reptiles and higher vertebrates, including humans.

---

not deserve to have the theory of strict recapitulation, involving terminal addition, attributed to him. Gould (1977) attributes certain merit to Haeckel's biology, although he was keen to distance himself from Haeckel's presumed ideas on eugenics.

When it comes to language, Anderson and Lightfoot (2002) agree that we can have a deterministic theory of language acquisition, where language seems to 'grow' in the human infant from simple one-word communication right up until fully grammatical sentences are produced. There appears to be a 'trend' toward complexity as the human brain grows dramatically in size and connectivity. However, we cannot assume that this seemingly biologically determined system of language acquisition can be equated with a deterministic history of language change, both within our modern human ancestors or especially within our hominid ancestors.

Moreover, Lock (1993) argues that there are no real stages of language acquisition, despite appearances, but we refer to a "one-word-stage" purely for convenience. Rather, we see different abilities "merge" from one into the next. Words are not phonetically organized like an adult vocabulary, according to Lock, but are instead non-grammatical one-word, or two-word, sound referents to objects or events like "baby cry", which are learned by rote. He is thus cautious about a direct reading of human development expecting to find the evolutionary pathway to language. Theories of evolution that put forward assumed behaviour of adult hominid ancestors, and expect to find comparisons to the ontogenetic pathway of modern human infants, are for Lock, untenable.

In fact, as Vihman and DePaolis (2000) have pointed out, we find a completely different developmental pathway between human infants and apes. Human infants up to 6 months old seem to share a primitive type of episodic processing with that ascribed to ape culture, which is unreflective and bound to the immediate situation. However along with this simple episodic memory, human infants also have an instinctive sense of self that is lacking in other non-human primate infants. It is most unlikely that we would find any simple stage-like links between ontogeny and phylogeny of cognitive stages between humans and our primate ancestors, including hominids.

Chomsky (2005) supports Lock's findings that during acquisition of language children do not pass through a two-word state, etc., but incorporate their productions within a natural ability to use *unbounded Merge* with different interface properties, including semantics. Chomsky favours the primacy of semantics in language design and believes that the computational efficiency apparent in language communication is there to satisfy the semantic interface. He is, however, skeptical of accounts that posit an earlier stage of language that was just an independent language of thought. Rather, he supports the view that the origin of language was more likely the result of a genetic event that rewired the brain of *H. sapiens.* Chomsky notes the sudden and emergent "great leap forward" in expression of symbolic thought through language and symbolic behaviour, as revealed by the archaeological record.

It is important to note that much of the seemingly unique traits of cognition appear to emerge postnataly in humans. At the age of two years, a second cycle of postnatal development begins in the human infant, whereby neural circuits in and around Broca's area develop and allow for the hierarchical assembly of language skills (Pollack, 1994). This second cycle is lacking in other primates. Chimpanzees have a Broca's area, albeit much smaller than a human's, but there are no connections to other frontal regions of the brain as we see in modern humans. Moreover, it has recently been shown that the presence of a Broca's region does not automatically imply the capacity for speech. Walker and Shipman (1996) point out that PET scans of brains show that Broca's area is not just for speech, but is more for the coordination of complex motor actions. In fact, Broca's area is not always activated during speech, so the evidence showing that *H. erectus* had a Broca's cap has mistakenly led to the claim that this species must have had speech. Lesions in Broca's area may produce stuttering and various other motor problems related to speaking, adding further evidence that Broca's area is mostly involved in the *production* of speech, not necessarily in the motivational cognitive processes involved in the generation of speech (Pollack, 1994).

It is entirely likely that the expansion of Broca's area is directly related to the apparent enhancement of manual dexterity in *H. erectus*[13] rather than language skills. Lieberman et al. (1991) remind us that Broca's area is not solely dedicated to language, but is involved in precise, lateralized hand movements. A larger Broca's area aiding an increase in manual dexterity may have offered a pre-adaptation for the motor functions of language. Bowers (2006) believes that the emergence of bipedalism may have had immediate developmental consequences. Areas in the sensory and motor 'homunculi' of the brain that were previously devoted to the grasping ability of the foot were likely to have been co-opted for manual dexterity. Bowers notes that the motor and sensory representational areas in the brain are largely set up as a result of neural inputs after birth[14].

Friederici (2006) has shown that the brain basis for language continues over time as the brain matures, and the processes that develop in the infant function the same as in the human adult brain. Lateralization of the human brain is present at birth. Phonological processes are acquired during the first months, followed by semantic processes at 12 months. Syntactic structures are acquired continuously in the second and third year, and by 30 months, syntactic processes with all of their constraints have emerged. Friederici views the developmental process of human language acquisition from infant to adult as more quantitative than qualitative as the neural basis for language is already present in early infancy.

For Jackendoff, a highly modularised language processor has evolved in incremental parcels, which have operated at different times in parallel with each other. He argues that language has evolved like biological species and says that "cases sometimes

---

[13]Both *Homo habilis* and *H. erectus* appear to have had more human-like fingertips and thumbs that would have allowed for a precision grip (Ambrose, 2001).

[14]Bowers (2006) notes that areas in the brain devoted to the hands and wrists are typically larger in individuals who trained on the violin from early childhood.

arise where what is apparently one species shades off impercep-
tibly over some geographical range into another" (Jackendoff,
2002). This 'shading off' is actually *not* what we find in nature.
These sorts of arguments that appeal to commonly held, but
false, views about how evolution of species proceeds, further
diminish the credibility of the gradualist approach.

Taken together, all of the arguments for recapitulation of language evolution in infant language acquisition do not hold up.
Our hominid ancestors experienced different life-histories and
brain growth trajectories. There is no evidence for a gradual
increase in brain growth, or terminal addition during the developmental phase of our hominid ancestors. Brain growth
trajectory, and brain size to body size ratio in pre-human hominids, have been shown to be similar to extant apes. We
have no reason to believe that an 'ontogeny' of language in
our hominid ancestors followed any of the 'stages' of human
language acquisition. In fact, we find none of the cognitive
traits that appear to be unique to humans. The archaeological
record supports this conclusion.

7.7   Language as Exaptation

Exaptation, a term invented by Gould and Vrba (1982), is characterised as the appearance of a new trait, which has derived its adaptability from a previously evolved form. There are many supportive functions of language that are not associated with cognition, but appear to have been adopted (not adapted) or co-opted for speech or sign. Some of these features are purely anatomical. For example, we have seen in chapter 4 that the human spine has a much expanded amount of gray matter over that of other primates. Gray matter contains nerve cell bodies important for the innervation of motor functions controlling the thoracic and abdominal muscles implicated in the fine control over breathing in addition to fine motor control with the hands (MacLarnon and Hewitt, 2004). This modification has allowed for the fine control of breathing during speech.

Earlier, I put forward evidence that a simple change in the developmental pathway of our immediate *H. erectus* ancestor has caused the apparent paedomorphic anatomy of the human face and skull. Nishimura et al. (2006) have found that the neotenous retention of a flat face, as we find in fossil *H. erectus* infants, has modified the human supralaryngeal vocal tract (SVT)[15]. The enhanced breathing afforded by the newly modified spinal column has allowed humans to take advantage of the newly evolved SVT. Humans are able to finely modulate sound due to the configuration of our long pharynx, or upper throat, tongue, and lips. The pharynx is much longer than other primates due to the descent of the larynx to a position

---

[15]In all primates, the mouth bifurcates into two tubes, one for breathing and the other to carry food into the stomach (Pollack, 1994). In mammals, the larynx is high in the throat and locks the air space at the back of the nasal cavity (nasopharynx). This configuration allows for continued breathing and swallowing of food at the same time, but also limits the range of sounds that mammals can make, due to the reduced pharyngeal cavity necessary for sound amplification. This typical mammalian anatomical pattern is present in human newborns, which means that they can suckle and breath at the same time. As humans develop, they lose this ability.

lower in the neck. Our fine control of breathing allows for the wide variety of output, from punctuated bursts to smooth continuous sounds (MacLarnon and Hewitt, 2004).

The exaptation approach goes against the often received view that the human vocal tract must have gradually evolved to support and improve speech. For example, Bickerton states that

> it is easy to see how increasing numbers of words and/or increasing ability to combine these words into longer and more complex sentences would have forcefully selected for improvements in the vocal organs and increasing complexity in the categorization of sounds (Bickerton, 2007).

Similarly, Pinker (1994) thinks that the human vocal tract is tailored to the demands of speech, under the assumption that language evolved in order to 'enhance' communication.

As argued previously, we should not point to the current utility of a biological trait in order to infer that the trait has evolved *because* it is now useful. Useful traits, like the human vocal tract, have been exapted for speech, rather than having been modified or adapted to *serve* this mode of language production. In other words, the SVT constrains the possible articulations of language, rather than having been 'tailored' to suit the requirements of speech (Piattelli-Palmarini, 1989; Jenkins, 2000). The emergent and plastic quality of speech is supported by Kelso (1995), who points out that the physiology that enables speech is the result of self-organizing systems involving the coordination of the brain, the respiratory and laryngeal systems, and the oral cavity. The utterance of a single syllable requires the coordination of approximately 36 muscles, operating as a functional synergy. To illustrate the plasticity of this system, Kelso encourages us to try an experiment, whereby we disrupt this system by putting a pen in the mouth, disrupting jaw movement. It makes no difference because the overall functionality is preserved due the brain's ability to self-reorganize and dynamically stabilize the system. It is most likely that all of the physiological support systems that enable speech were exapted by the brain to use as a communication device. Rather than having evolved separately and gradually, the fossil evidence points to a uniqueness of these anatomical modifications in *H. sapiens*[16] although they have been co-opted from previously evolved systems unrelated to language.

Other cognitive traits can be identified that appear to have precursors in our primate brain. Our newly evolved cerebral cortex, according to Panksepp (1998), provides ongoing inhibition over ancient subcortical processes. He notes that a neocerebellar cortex has also evolved in the human brain, and this exerts a similar effect on the deep nuclei of the cerebellar (little brain), which is the seat of violent impulses. It is interesting to note that our 'seeking' mechanisms, which create our affective states of expectation and hence frustration, are closely linked with the primate rage circuits. Panksepp has suggested that our internal dialogs that apportion blame and scorn, emerge from the ancient 'rage' circuits shared by all mammalian brains.

Staying with the theme of descent with modification, Marcus (2006) believes that we should look to other *cognitive* domains that may underlie, and be the precursors for, our language faculty. For example, language relies heavily on memory, just like many other cognitive aspects. He argues that humans may have evolved a special mechanism for the encoding, storage and retrieval of language data, which overlaps with the retrieval of memory in other cognitive domains. He adds that the basic system for memory retrieval may be shared across the vertebrate world. The human brain has experienced a dramatic increase in white matter, which affords greater connectivity, and has allowed for an increase of processing temporal information. An increase in the anterior portion of Broca's area has implications for human language, especially for semantic

---

[16]See chapter 4 for further evidence supporting this conclusion.

information (Schoenemann et al., 2005). Schenker et al. (2005) have employed comprehensive imaging studies of both human brains and those of rhesus monkeys. They demonstrate that the frontal lobe comprises anatomical subdivisions with distinct functional attributes like perception, response selection, working memory, and problem solving. The enlarged gyral white matter within any given brain region indicates increased interconnectivity both within and between neighboring cortical regions as well as intrahemispheric and interhemispheric connections in the corpus callosum. We should consider this modification of the human brain as exaptation of existing neural systems, but this modification has led to quantitative as well as qualitative enhancements.

Likewise, the machinery for sequencing linguistic material may be an extension of non-cognitive sequencing tasks. Our underlying ability to represent space and time may have been co-opted for language, which contains many metaphors for space and time (Marcus, 2006). Motor planning, according to Marcus, is a further cognitive skill that may have been borrowed for language processing because it shares similar functionality of structured, hierarchical relationships.

I have already touched on the work on mirror neurons. Arbib (2005) believes that mirror neurons are important for the ability to imitate, and may have provided the scaffolding for the emergence of proto-language. Kaplan and Iacoboni (2005) have shown that some mirror neurons in human subjects respond equally to the sight *and* the sound associated with an action. They suggest that the ability of the motor cortex to represent both modalities may have been the key for the emergence of true symbolic capability. Motor activation to action sounds is typically lateralized to the left hemisphere, whereas sight of an action is typically bilateral. Monkeys do not show this same left lateralization to action sounds, and it is believed that this may have been the evolutionary change that made human brain functions such as language possible. However Indurkhya (2005) is dismissive of the mirror neuron as a crucial factor in the evolution of language. Monkeys merely produce behaviour that emulates another's behaviour based on imprinting or stimulus enhancement, they do not imitate. The empathetic nature of the human experience – being able to project one's thoughts and feelings into another person – is for Indurkhya (2005) the crucial distinction between humans and other animals. It is unlikely that early hominids had this ability to interpret a wide variety of actions and situations. Only with _H. sapiens_ do we find evidence for ritual behaviour indicative of the capacity for empathy (e.g., burial and artifacts of adornment).

Many primates have Broca's areas that appear to function like that of a human one-year-old (Pollack, 1994). The Broca's area in infant and adult chimpanzees, and humans up to about the age of two, serves complex hierarchical manipulations. At the age of two years, the Broca's area is larger in humans than in the chimpanzee and begins a second cycle of postnatal development, whereby neural circuits in and around the Broca's area makes connections to other frontal regions in the brain. It appears that this second cycle of neural connections correlates with the rapid onset of language skills attained by human infants. This second cycle is completely lacking in other primates. Pollack observes that other primates may have a rich vocabulary of sounds, but do not have the ability, like humans, to arrange them into a hierarchy of complexity. Damage to the human Broca's area can cause loss of the ability to maintain a hierarchical structure, but leaves the basic ability to string words together intact.

Cook (2002) points out that the right hemisphere, although it does not control speech, is essential for its interaction with the left hemisphere for normal language processing of small units like morphemes and phonemes as well as utterances like sentences and jokes. Hemispheres appear to become specialized during development and as Cook points out, may take over the role of the opposite hemisphere if that hemisphere is damaged in infancy. Our hominid ancestors show some brain asymmetry, but other living primates also show brain asymmetries, so, as Steele (2002) points out, this is not a uniquely derived human trait that can be linked to language. For Steele, brain asymmetries are just an emergent property of larger primate brains, and cannot be linked to handedness nor language ability. Great Apes show no tendency for dominance of the right arm, in fact, no evidence of hand preference at all, even with an asymmetric brain.

On the other hand, Crow (2005) thinks that an additional factor is involved in brain asymmetry. The human brain shows a difference in distribution of tissue within the two hemispheres creating the cerebral torque, and this difference allows the spread of neural activity. He thinks that the differing connections of commissural fibres between the two hemispheres may underpin the emergence of the ability to share and direct attention in triadic communication. We saw in sections 7.2.3 and 7.2.4, how understanding the intentions of others is foundational for deciding what the other is trying to communicate, and its importance for language. Crow (2006) believes that a single gene may account for the asymmetry of the human brain and may be a major determinant of cognitive ability. It appears that this single gene change has allowed for a different trajectory of brain growth in humans, and can be deemed epigenetic in nature, rather than following a strict blueprint of development. His work with schizophrenia has shown a definite link between this disorder and syntax, and is related to variations in asymmetries during brain development.

7.7.1 Language *for* Thought

Christiansen and Chater (2008) believe that there are many pre-adaptations in the brain, like the ability to reason about other minds, to represent discrete symbols, to understand and share intentions, and other biological traits such as increased memory and a prolonged childhood. They believe that changes in the brain have provided the underlying capacities for these sharing practices and may have shaped our cultural traditions. They doubt that our language faculty could have developed under a traditional Darwinian, adaptationist model. For them, language does not have a specialized biological machinery, but has evolved to fit with other learning and processing constraints within the brain. Their approach is cognition first, then language, which is in direct contrast to Schoenemann (2006), who finds it "difficult to escape the conclusion that language likely played a major role in the evolution of the human brain".

Bickerton (2005b) takes a slightly different approach. He is skeptical about Chomsky's (2005) crucial rewiring of the human brain hypothesis, and believes that language and thought must have co-evolved once a protolanguage emerged. It is inconceivable how this co-evolution could have taken place and a totally inconsistent notion. Regardless of whether thought evolved before externalization in the form of language, Bickerton insists that there must have been some sort of selectional agency. Once again, it is difficult to justify this position as many other primate species would have been subject to the same 'selection pressures' as our hominid ancestors.

Taking a strictly internalist view of the underlying complexity of the linguistic mind, Hinzen and Uriagereka (2006) believe that the hierarchical, or multi-dimensional, nature of semantics and syntax may be constrained by our mathematical capacity, although all three systems are highly correlated. They ponder the question of how evolution could explain the "unexplained universality" of this correspondence. However, Piattelli-Palmarini (1989) is not so sure about the link between logic and language. He believes that all natural human languages show hardly any instance of what we might think of as communicative efficiency or constraints dictated by the laws of pure logic. He thinks that languages are quite peculiar in that they show little in common with artificial languages or formal logic, and that they therefore are unlikely to be linked to our seemingly natural ability for simple logic. Moreover, the deductive powers that a human infant brings to language acquisition are totally unlike the theorems of mathematics and logic, and cannot be formulated under any elegant learning theory. Piattelli-Palmarini concludes that trying to subsume language acquisition and language evolution under categories of general intelligence, communicative function, or problem solving, is totally useless.

7.7.2 Language only with *H. sapiens*

In chapter 4, I presented evidence from the hominid fossil record supporting the view that anatomically modern *H. sapiens* arose from a speciation 'event'. Lieberman et al. (2002) have shown how just a simple developmental change can induce the facial retraction and neurocranial globularity of the human skull. Whether this change was induced by a shift in cranial base angle, or a change in growth patterns of the human frontal and temporal lobes, is a question they leave open. They nevertheless support the view that speciation can result from ontogenetically early alterations in the regulation of growth leading to novel phenotypes. A simple morphogenetic change can induce a cascade of developmental changes that are highly integrated during the growth of the skull. It seems that a simple tinkering with the developmental program of immediate ancestor, *H. erectus* (or a subspecies thereof), produced a paedomorphic species, *H. sapiens*, with all of the anatomical pre-adaptations for language (Gould, 2000; Arsuaga et al., 1999b; Nishimura et al., 2006).

Twenty years ago, Chomsky (1988) was confident that molecular biology would have more to say about evolutionary theory than the theory of natural selection, which has little to say about how mutations arise. Rosselló and Martín (2006) believe that Chomsky's more parsimonious approach with his Minimalist Program, may actually be far more suitable for evolutionary inquiries. They support Chomsky's (2005) suggestion that we should consider the possibility of a discontinuity for the emergence of language when a certain degree of neural complexity is attained. In chapter 6, I elaborated on the changes that appear to have taken place in the human brain (Araque, 2008; Cáceres et al., 2003; Hayakawa et al., 2005; Rae et al., 2003; Rilling and Seligman, 2002), which have given us our expanded cortical plate, and the enhanced patterns of connectivity to phylogenetically older structures. I have argued that a simple genetic change in the human brain has led to the sudden emergence of complexity in cognition that is not found in any other species, and, according to the archaeological record, not in any of our hominid ancestors.

7.8   Summary

In this chapter, I have argued that the gradualist approach to the evolution of language *for* communication is not feasible. It has been shown that the precursors for human language, including shared intensional thought, are totally lacking in all other primates. In addition, human speech or gesture is fundamentally different from animal calls, including those of chimpanzees. We cannot point to any obvious differences in social structure between any of the Great Apes and our hominid ancestors. So, arguments based on selection pressures resulting from the 'need' to communicate within a complex social organization are very difficult to defend.

Bickerton (1998), using the analogy with the pidgin to creole transition, which can occur within a single generation, argues that fully syntactic language may have arisen similarly in one major leap. I have surveyed the many theories put forward for how a protolanguage may have emerged, including Bickerton's foraging theory, but I find that in the main they are all a case of special pleading. Nevertheless, his point that fully syntactic language can emerge suddenly is further support for the saltatory evolution of language. He raises a reasonable objection against the gradualist program for the evolution of language modules. Today's languages either have syntax or they do not. Syntaxless pidgin 'languages' can transform into fully syntactic creoles in a single generation. If syntax had developed gradually, with the addition of "quasi-independent" modules, then we would expect to find some sort of deficit today if one of these syntactic modules failed. Bickerton points out that no such deficit can be linked directly to any specific syntactic aspect. This suggests that syntax is a single module and may have emerged in one single step in the same way that creoles can emerge from pidgins in one generation.

Bickerton (2002) has argued that the ancestral hominid brain was built to attend to 'stuff' coming in from outside, not to 'stuff' coming from the inside. In this sense, the brains of ancestral hominids were probably more like other mammals, that is, dealing with a world that was here and now rather than reflecting on a past or possible futures. This scenario is supported by the archaeological record of hominids prior to *Homo sapiens*. With the sudden appearance of the enlarged neo-cortex in *H. sapiens* and the apparent massive increase in memory, the human brain had to cope with 'stuff' coming from the *inside*.

Lightfoot (2000) agrees that evolution is typically a discontinuous process. Only those innovations which are big enough and effective enough to be adaptive will make a difference. He is more sympathetic to the view that 'explosive' changes can cause qualitative modifications in organisms, rather than having evolved gradually. This is a particularly important point when evoking change in the language faculty, where transitional forms of elements of grammar do not make sense. For example, how could we have a halfway position between the subject-verb-object language and a subject-object-verb language? (Anderson and Lightfoot, 2002).

The argument for a single-step evolution of human cognition and language is supported by the sudden explosion of complex behaviour and technology that coincides with the emergence of fully anatomically modern humans in southern Africa around 120,000 years ago. Bickerton (2002) notes that this avalanche of technological advancement, and apparent cognitive complexity, eventuated after a stasis of about 2 million years. It is inconceivable that hominid cognition, including all of the traits that are foundational for language, would have evolved gradually over this vast amount of time without leaving any evidence in the archaeological record.

8. Conclusion

I set out to show that the saltationist stance is the only defensible approach toward the evolution of humans and the unique aspects of human cognition. When we consider the vast periods of time since the split from our Great Ape ancestor, and the non-existent footprint that hominids left on the landscape for the first 3-5 millions of years of 'evolution', we can only conclude that human-like cognition did not evolve in a gradual and adaptive pattern during this time.

The evidence from paleontology and archaeology is consistent with a picture of sudden emergence of the first hominids, followed by many millions of years of stasis. The earliest hominids have been dated to around 7 millions years ago, but the first 'signs' of any advance in cognitive ability emerge only around 2.6 million years ago (Semaw, 2000). I have chosen to put the term 'signs' in scare quotes due to the fact that I believe that the simple, single strike stone tools that are found at this time show little, if any, evidence for a capability that is beyond that of extant chimpanzees. Several workers, as we saw in chapter 5, have supported this viewpoint. Gargett (1993) attributes this so-called stone tool 'technology' to "breaking rocks" and Mercader et al. (2002) have shown that many of the abilities attributed to the makers of these tools can be found in other Great Apes.

Our presumed immediate forebear (*H. erectus*) made a stone tool that is considered by some to represent a major advance in cognition, that is, the ability to plan ahead and 'construct' a tool from a pre-conceived mental template. The Acheulian hand-axe was a bifacially worked stone tool that in some cases

took on a symmetrical shape. This shape has led many scholars
to assume that the symmetry must have been a standardized
form that the stone knapper had 'in mind'. The end product
may have been useful as a chopping or scraping tool[1], but many
scholars have argued that the sometimes symmetrical shape
of this stone tool was just the end product of the reduction
process[2]. The final 'hand-axe' appearance was determined by
the size, shape, and raw material (rock) used. There is no
doubt that its makers appear to have had a greater control
over the production process. However, many believe that the
greater dexterity of these tool-makers can be accounted for by
rather simple, although beneficial, changes in the shoulders,
wrists, and fingers that are apparent with these later hominids[3].
Evidence of planning and foresight is claimed by some scholars
because it appears that some of the raw material used in the
making of these stone tools was deliberately transported to
knapping sites. However, chimpanzees have been known to
transport raw materials in order to crack panda nuts, so this
assumption seems unwarranted (Mercader et al., 2002).

I was also critical of many selectionist theories purporting to
have the answers for how and why our hominid ancestors had
gradually evolved away from their Great Ape predecessor. It
has been shown that hominids occupied the same sort of terri-

---

[1]Although it is interesting that analysis of use-wear on these stone tools,
which were prolific in the archaeological landscape, has not definitively
revealed what these stone tools were actually used for – they may just be
the left over cores from which flakes have been removed.

[2]I like to use the analogy of the reduction process when one eats an
apple. The resulting core usually ends up being symmetrical, but we can
easily see that this symmetry results from the practicalities of the reduction
process.

[3]There is the possibility that greater dexterity arose through the process of natural selection, but I believe that it is more plausible that this
'developmental set' arose through a heterochronic change during embryonic
development. Hlusko (2004) points out that modifications in the pelvic
region, the feet and the hands in *Australopithecus* were due to a mutation
that affected the developmentally functionally interrelated set of correlated
traits. The traits enabling bipedality were linked with the development of
hands with shorter fingers.

tory that other primates inhabited. We have no cause to assume that hominids were living under different ecological 'pressures' than those of extant chimpanzees. Hominids were not 'forced' into living on the savannah by environmental change, nor were they 'forced' into living in collaborative social structures that 'required' the acquisition of communicative systems. I repeat the apt sentiment from Leakey (1992) regarding the adaptationist notion, that somehow humanity made the required initiative and effort to evolve a more complex lifestyle, whereas the apes remained apes because they didn't exert themselves enough. The more plausible explanation behind hominids moving out of the trees and onto the savannah comes from developmental biology, where it has been shown how a rather simple mutation at the stage of pelvis and limb development produced a bipedal ape in one saltational event. This hominid, although retaining some ability for tree climbing, was most likely more agile with its newly evolved locomotive ability. The most compelling theory for the sudden emergence of hominids comes from Bowers (2006) who points out that the HoxD sequence, which controls the formation of the vertebral column, pelvis, and limbs, resides on the human chromosome 2. Chromosome 2 resulted from the mutation that caused the reduction in the number of chromosomes from the pongid 48 to the human 46, and is believed by Bowers to be the quintessential mutation that led to a dramatic change in the functionality of the locomotive system.

Apart from a few unsophisticated stone tools, our hominid ancestors appear not to have made any cognitive advance over that of extant Great Apes. Before the emergence of *H. sapiens*, our ancestors did not engage in complex collaborative activities like hunting, preparing and sharing meals, and building shelters. They did not engage in any rituals that celebrate various life-stages, the most visible that we recognize as human-like being burial of the dead. The archaeological record has not brought forward any evidence of tokens relating to art, personal adornment, or items of any symbolic nature. These collective activities together with a sense of an ego are underpinned by social 'institutions' in modern humans, but we have no evidence of any social systems at all associated with our hominid ancestors. Only with the sudden emergence of *H. sapiens* do we find artefacts that might reflect something akin to human-like cognitive behaviour. As more archaeological sites are discovered in Africa, the emergence of symbolic artefacts is pushed further back in time toward a confluence with the emergence of modern human anatomy, around 120,000 years ago.

I have contended that modern humans (*H. sapiens*) arose as the result of a single mutation that radically altered the developmental program of our immediate forebear. We are the neotenous descendants of *H. erectus*, most likely of African origin. Genetic evidence points to a recent emergence of *H. sapiens* in Africa around 120,000 years ago (Adcock et al., 2001; Drayna, 2005; Mitteroecker et al., 2004; Pearson, 2004). Although sparse, fossils of *H. erectus*[4] infants and juveniles provide compelling evidence for this claim. Humans have retained the globular skull and the flat face of a *H. erectus* infant. Our teeth are much smaller, which is consistent with a pattern of a slower development. In fact, neotenous retention of a flat face has altered our whole orofacial design, and has modified the human supralaryngeal vocal tract (Dean, 2006; Nishimura et al., 2006). The retention of the globular skull of an infant has allowed room for a change of development for the human frontal and temporal lobes, essential for the hallmarks of human cognition like creative thinking, artistic expression, planning and language. Other modifications of the nervous system appear to have accompanied this major mutation. The spinal column in humans has greatly enlarged in comparison with adult *H. erectus*. This particular change has allowed for the enhancement of many human behaviours like coordinated running and walking, fine motor control over the hands for manipulation and production of complex artefacts, and breathing control important for spoken language (MacLarnon and Hewitt, 2004).

---

[4]I have pointed out previously that there are claims for many subspecies of *Homo* e.g. *H. heidelbergensis*, *H. helmei*, *H. rhodesiensis*, and any one of these populations could have been our immediate ancestor.

Although we are a long way from identifying the crucial differences between human and ape-like cognition, we nevertheless are able to point to some of the radical differences in the building of the human brain during development. Ramus (2006) believes changes to regulatory systems of gene expression have been a major factor in determining the human phenotype. He finds that about 70% of our gene variants are human specific, and a large proportion are expressed most highly in the brain. I have outlined some of the recent findings from developmental neurobiology that offer tantalizing suggestions for how a radical change in human brain architecture has allowed us to escape the restraints on cognitive behaviour that is apparent in all other animals. Rakic (1995) contends that the mutation of regulatory genes that control the timing and ratio of cell divisions, which create the initial proliferative cells (stem cells) before neurogenesis, has established new patterns of connectivity and an expanded cortical plate in the human brain. It appears that a delay in the onset of the second phase of corticogenesis by a few days allows for three to four extra rounds of mitotic division, thereby producing a massive increase in the number of founder cells, which in turn produce the expanded cortical surface. The species-specific size of the cortex is determined by the pool of proliferative cells in the early stages of embryonic development.

Cáceres et al. (2003) find that approximately 90% of the genes that are involved in building the primate brain are more highly expressed in humans. The differences between the species result not from a difference in gene sequences, but primarily from alterations in their expression. This forms the basis for extensive modifications of cerebral physiology and function in humans. It is these biochemical changes in human neural cells that enable them to function longer than those in other primates, leading to high levels of cerebral activity over the lot longer life span in humans. Humans live on average for around 40-50 years longer than the Great Apes and also, it appears, their hominid ancestors (Stringer and Gamble, 1993). An important difference in brain microglia between humans and chimpanzees is possibly due to changes in regulatory sequences of Siglec-11, a novel human molecule prominently expressed in the human brain (Hayakawa et al., 2005). This gene duplication, which is universal to modern humans, is a recent gene conversion event. The extended life-history of humans, in line with an extended viability of cerebral activity with the newly acquired age group of care-givers, undoubtedly afforded a great advantage to early humans.

As we saw in chapter 6, other important changes in human brain architecture have been brought about by the significantly larger amounts of white matter (Rilling and Seligman, 2002; Schenker et al., 2005; Schoenemann et al., 2005). The proportion of white matter volume can be as much as 80% larger than expected for primates. The disproportionate size of the human temporal lobe white matter augments the number of connections linking temporal and prefrontal cortex. This novel architecture is important for the mediation of visual processing, auditory processing, social behaviour, and association cortex, all implicitly involved in the production and comprehension of human language.

Astrocytes, which make up most of the white matter in the brain, are an important and necessary support for neuronal activity, providing not only metabolic support for neurons, but also playing an important information processing role in the central nervous system (Araque, 2008). The human brain has an unusually high metabolism, which is supported by Astrocytes, which in turn support the energy requirements for neurons (Cáceres et al., 2003). Astrocytes are involved in the glutamate/glutamine cycle in the brain enabling a fast, energy efficient recycling of neurotransmitter glutamate, and circulating glucose concentrations have very broad and robust influences on brain functions, enhancing learning and memory (Gold, 1995; Rae et al., 2003). In addition, Astrocytes, like neurons, are able to sense synaptic activity as an input signal, and are able to integrate these signals into output signals, thereby actively modulating neuronal excitability and synaptic transmission (Araque, 2008).

The expansion of white matter, particularly in the anterior portion of Broca's area, has implications for human language, especially semantic information. Schoenemann et al. (2005) note that the greater interconnectivity between the prefrontal cortex and many other brain areas, afforded by the dramatic increase in white matter, have allowed for an enhancement of processing of temporal information, which in turn has given humans an understanding of causality. An understanding of causality, they believe, is based on the ability to remember the temporal order of past events, and recognizing causal relationships underpins general-problem solving abilities – critical for the development of elaborate technology. As we saw in chapter 7, an understanding of causality has also been important in the process of language acquisition, according to many linguists (e.g., Brownell et al., 2005; Lock, 1993; Markson and Diesendruck, 2005; Watson, 2005).

I have argued throughout that our hominid ancestors did not possess any of the cognitive abilities that appear to be unique to humans. We can see from the fossil record that *H. erectus* brain growth followed a similar trajectory to other Great Apes. Hominid brain size, before the emergence of *H. sapiens*, was unremarkable in that it correlated with what we would expect for a primate of that body size (Ankel-Simons, 2000; Eldredge, 1996; Ruff et al., 1997). In addition, it has been shown that *H. erectus*, although its absolute brain size may have been similar at birth to that of humans, did not follow the same developmental curves (Hublin and Coqueugniot, 2005). In fact, a 1.8 million year old *H. erectus* specimen, aged between 0.5-1.5 years, appeared to have followed a brain growth pattern similar to that of a chimpanzee. This pattern, as evidenced by the fossil record, did not change for the entire period of *H. erectus* evolution (Antón, 2004; Rightmire, 1990; Ruff et al., 1997). Chimpanzees achieve approximately 80% of adult brain size by one year old and it appears that *H. erectus* brain growth followed the same pattern. The *H. erectus* brain had also almost reached its adult size by the end of the first year after birth, and we must remember that it was only the size of a brain belonging to a one year old human. The human brain reaches about 825 grams by the end of the first year, which is roughly the size of a fully developed *H. erectus* brain (Montagu, 1989). It is therefore unlikely that the pattern of *H. erectus* brain growth allowed for the important cognitive traits that we find emerging in humans from infancy and persisting throughout the juvenile period, laying the foundation for ongoing cognitive ability in adulthood. Their fast growing brain appears to have closed the window on the ability of *H. erectus* to acquire a human-like complexity in cognitive behaviour, and most likely, to acquire language.

It is highly unlikely that the complexity of human cognition arose in a gradual and adaptational pattern. The unique traits commonly assigned to human cognition, including language, emerged in a saltational event, along with fully modern human anatomy, in Africa, around 120,000 years ago. The archaeological evidence, albeit scarce, supports my claim. No matter how unpalatable saltationism might be to Darwinian gradualists, in making my case for a saltational approach for the evolution of human cognition and language, I chose to go with the evidence, rather than the majority view.

Bibliography

Ackermann, R. R., Rogers, J., and Cheverud, J. M. (2006). Identifying the morphological signatures of hybridization in primate and human evolution. *Journal of Human Evolution*, 51:632–645.

Adcock, G. J., Dennis, E. S., Easteal, S., Huttley, G. A., Jermiin, L. S., Peacock, W. J., and Thorne, A. (2001). Mitochondrial DNA Sequences in Ancient Australians: Implications for Modern Human Origins. *PNAS*, 98:537–542.

Alba, D. M., Moyà-Solà, S., and Köhler, M. (2001). Canine reduction in the Miocene hominoid *Oreopithecus bambolii*: behavioural and evolutionary implications. *Journal of Human Evolution*, 40:1–16.

Ambrose, S. H. (2001). Paleolithic Technology and Human Evolution. In Scher and Rauscher (2003), pages 1746–1753.

Amunts, K., Lenzen, M., Friederici, A. D., Schleicher, A., Morosan, P., Palomero-Gallagher, N., and Zilles, K. (2010). Broca's Region: Novel Organizational Principles and Multiple Receptor Mapping. *PLoS Biology*, 8 (9) doi:10.1371/journal.pbio.10000489.

Anderson, S. R. and Lightfoot, D. W. (2002). *The Language Organ*. Cambridge University Press, UK.

Ankel-Simons, F. (2000). *Primate Anatomy*. Academic Press, USA.

Antón, S. C. (2002). *Cranial Growth in Homo Erectus*, pages 349–380. In Minugh-Purvis and McNamara (2002).

Antón, S. C. (2004). The face of Olduvai Hominid 12. *Journal of Human Evolution*, 46:337–347.

Anwander, A., Tittgemeyer, M., von Cramon, D. Y., Friederici, A. D., and Knösche, T. R. (2007). Connectivity-Based Parcellation of Broca's Area. *Cerebral Cortex*, 17:816–825.

Araque, A. (2008). Astrocytes process synaptic information. *Neuron Glia Biology*, 4:3–10.

Arbib, M. A. (2005). From monkey-like action recognition to human language: An evolutionary framework for neurolinguistics. *Behavioral and Brain Sciences*, 28:105–167.

Archer, J. (1998). *The Nature of Grief*. Routledge.

Archer, M. (1988). Our Oral Arsenal: A Natural Legacy of Primate Aggression. *Australian Natural History*, 22:474–475.

Argue, D., Donlon, D., Groves, C., and Wright, R. (2006). *Homo floresiensis*: Microcephalic, pygmoid, *Australopithecus*, or *Homo*? *Journal of Human Evolution*, 51:360–374.

Armstrong, D. F., Stokoe, W. C., and Wilcox, S. E. (1994). Signs of the Origin of Syntax. *Current Anthropology*, 35:349–368.

Arnason, U., Gullberg, A., Burguete, A. S., and Janke, A. (2000). Molecular Estimates of Primate Divergences and New Hypotheses for Primate Dispersal and the Origins of Modern Humans. *Hereditas*, 133(3):217–228.

Aronoff, M., Meir, I., Padden, C., and Sandler, W. (2008). Language is shaped by the body. *Behavioral and Brain Sciences*, 31:509–511.

Arsuaga, J., Lorenzo, C., Carretero, J., Gracia, A., Martinez, I., Gracia, N., de Castro, J. B., and Carbonell, E. (1999a). A Complete Human Pelvis from the Middle Pleistocene of Spain. *Nature*, 399:255–258.

Arsuaga, J., Martinez, I., Lorenzo, C., and Gracia, A. (1999b). The Human Cranial Remains from Gran Dolina Lower Pleistocene Site. *Journal of Human Evolution*, 37:431–457.

Baab, K. L. (2008). The taxonomic implications of cranial shape variation in *Homo erectus*. *Journal of Human Evolution*, 54:827–847.

Balzeau, A., Grimaud-Hervé, D., and Jacob, T. (2005). Internal cranial features of the Mojokerto child fossil (East Java, Indonesia). *Journal of Human Evolution*, 48:535–553.

Bara, B., Barsalou, L., and Bucciarelli, M. (2005). *Proceedings of the 27th Annual Meeting of the Cognitive Science Society*. Lawrence Erlbaum Associates, Mahwah, New Jersey.

Barrett, A. M., Foundas, A. L., and Heilman, K. M. (2005). Speech and gesture are mediated by independent systems: commentary on Arbib (2005). *Behavioral and Brain Sciences*, 28:125–126.

Barton, R. A. (2001). The coordinated structure of mosaic brain evolution. Reply to Finlay et al. 2001. *Behavioural and Brain Sciences*, 24:281–282.

Begun, D. R. (2006). Planet of the Apes. *Scientific American*, 16:4–13.

Bejder, L. and Hall, B. K. (2002). Limbs in whales and limblessness in other vertebrates: mechanisms of evolutionary and developmental transformation and loss. *Evolution & Development*, 4:445–458.

Bellomo, R. V. (1994). Methods of determining early hominid behavioral activities associated with the controlled use of fire at FxJj 20 Main, Koobi Fora, Kenya. *Journal of Human Evolution*, 27:173–195.

Bermúdez de Castro, J. M., Carbonell, E., Caceres, I., Diez, J. C., Fernandez-Jalvo, Y., Mosquera, M., Olle, A., Rodriguez, J., Rosas, A., Rosell, J., Sala, R., Verges, J. M., and Van der Made, J. (1999). The TD6 (Aurora Stratum) Hominid Site. Final Remarks and New Questions. *Journal of Human Evolution*, 37:695–700.

Bermúdez de Castro, J. M. and Nicolás, M. E. (1997). Palaeodemography of the Atapuerca-SH Middle Pleistocene hominid sample. *Journal of Human Evolution*, 33:333–355.

Bickerton, D. (1998). *Catastrophic Evolution: The Case for a Single Step from Protolanguage to Full Human Language*, chapter 21, pages 341–358. In Hurford et al. (1998).

Bickerton, D. (2002). *From Protolanguage to Language*, pages 103–120. In Crow (2002).

Bickerton, D. (2005a). Beyond the mirror neuron - the smoke neuron? : commentary on Arbib (2005). *Behavioral and Brain Sciences*, 28:126.

Bickerton, D. (2005b). Chomsky: Between a Stony Brook and a Hard Place. In *A report on Chomsky at the 2005 Stony Brook Symposium, S.U.N.Y.*

Bickerton, D. (2005c). Language first, then shared intentionality, then a beneficent spiral: commentary on Tomasello et al. (2005). *Behavioral and Brain Sciences*, 28:691–692.

Bickerton, D. (2007). Language evolution: A brief guide for linguists. *Lingua*, 117:510–526.

Binford, L. R. and Ho, C. K. (1985). Taphonomy at a Distance: Zhoukoudian, "The Cave Home of Beijing Man"? *Current Anthropology*, 26:413–442.

Bisson, M. S. (2001). Interview with a Neanderthal: An Experimental Approach for Reconstructing Scraper Production Rules, and their Implications for Imposed Form in Middle Palaeolithic Tools. *Cambridge Archaeological Journal*, 11:165–184.

Blomquist, G. E. (2009). Brief Communication: Methods of Sequence Heterochrony for Describing Modular Developmental Changes in Human Evolution. *American Journal of Physical Anthropology*, 138:231–238.

Blumenberg, B. (1983). The Evolution of the Advanced Hominid Brain. *Current Anthropology*, 24:589–623.

Bogin, B. (1999). *Patterns of Human Growth*. Cambridge University Press, UK.

Boisserie, J., Lihoreau, F., and Brunet, M. (2005). The position of Hippopotamidae within Cetartiodactyla. *Proceedings of the National Academy of Sciences of America*, 102:1537–1541.

Bolhuis, J. J. and Wynne, C. D. L. (2009). Can evolution explain how minds work? *Nature*, 458:832–833.

Botha, R. P. (2002). Did Language Evolve Like the Vertebrate Eye? *Language & Communication*, 22.

Bower, B. (1989). Talk of Ages. *Science News*, July 8:24–26.

Bowers, E. J. (2006). A New Model for the Origin of Bipedality. *Human Evolution*, 21:241–250.

Bridgeman, B. (2005). Action planning supplements mirror systems in language evolution: commentary on Arbib (2005). *Behavioral and Brain Sciences*, 28:129–130.

Brown, K. S., Marean, C. W., Herries, A. I. R., Jacons, Z., Tribolo, C., Brown, D., Roberts, D. L., Meyer, M. C., and Bernatchez, J. (2009). Fire as an Engineering Tool of Early Modern Humans. *Science*, 325:859–862.

Brown, P., Sutikna, T., Morwood, M. J., Soejono, R. P., Jatmiko, Wayhu Saptomo, E., and Due., R. A. (2004). A New Small-Bodied Hominin from the Late Pleistocene of Flores, Indonesia. *Nature*, 431:1055–1061.

Brownell, C. A., Nichols, S., and Svetlova, M. (2005). Early development of shared intentionality with peers: commentary on Tomasello et al. (2005). *Behavioral and Brain Sciences*, 28:693–694.

Brunet, M., Guy, F., Pilbeam, D., Mackaye, H. T., Likius, A., Ahounta, D., and Beauvilian, A. (2002). A New Hominid from the Upper Miocene of Chad, Central Africa. *Nature*, 418:145–151.

Buss, D. M. (2005). *The Handbook of Evolutionary Psychology*. New Jersey: Wiley.

Butterworth, G. and Franco, F. (1993). *Motor development: communication and cognition*, pages 153–165. In Kalverboer et al. (1993).

Byrne, L. (2004). Lithic tools from Arago cave, Tautavel (Pyrénées-Orientales, France): behavioural continuity or raw material determinism during the Middle Pleistocene. *Journal of Archaeological Science*, 31:351–364.

Cáceres, M., Lachuer, J., Zapala, M. A., Redmond, J. C., Kudo, L., Geschwind, D. H., Lockhart, D. J., Preuss, T. M., and Barlow, C. (2003). Elevated gene expression levels distinguish human from non-human primate brains. *PNAS*, 100:13030–13035.

Calvin, W. and Bickerton, D. (2000). *Lingua Ex Machina*. MIT Press, Cambridge, Mass.

Calvin, W. H. (2006). The Emergence of Intelligence. *Scientific American*, 16:84–92.

Cangelosi, A. (2000). Evolution of symbolisation in chimpanzees and neural nets. In *The Evolution of Language - 3rd conference. Paris, April 3rd - 6th, 2000*.

Cangelosi, A., Smith, A. D. M., and Smith, K. (2006). *Proceedings of the 6th International Conference (EVOLANG06) on the Evolution of Language, Rome*. World Scientific, London.

Caporael, L. (2003). *Repeated Assembly: Prospects for Saying What We Mean*, pages 71–89. In Scher and Rauscher (2003).

Carmody, D. P., Dunn, S. M., Boddie-Willis, A. S., DeMarco, J. K., and Lewis, M. (2004). A quantitative measure of myelination development in infants, using MR images. *Neuroradiology*, 46:781–786.

Carroll, S. B. (2006). *The Making of the Fittest: DNA and the ultimate forensic record of evolution*. W.W. Norton, NY.

Carter, R. (1998). *Mapping the Mind*. Phoenix, London.

Charman, T. (2005). Why do individuals with autism lack the motivation or capacity to share intentions?: commentary on Tomasello et al. (2005). *Behavioral and Brain Sciences*, 28:695–696.

Chase, P. G. (1990). Tool-Making Tools and Middle Paleolithic Behavior. *Current Anthropology*, 31:443–447.

Cheney, D. L. and Seyfarth, R. M. (2005). Constraints and preadaptations in the earliest stages of language evolution. *The Linguistic Review*, 22:135–159.

Chomsky, N. (1980). *Rules and Representations*. Columbia University Press, NY.

Chomsky, N. (1988). *Language and Problems of Knowledge*. MIT Press, Cambridge, Mass.

Chomsky, N. (2005). Some simple evo-devo theses: how true might they be for language? In *Morris Symposium on the Evolution of Language at S.U.N.Y. Stony Brook*.

Chouard, T. (2010). Evolution: Revenge of the hopeful monster. *Nature*, 463:964–867.

Christiansen, M. H. and Chater, N. (2008). Language as shaped by the brain. *Behavioral and Brain Sciences*, 31:489–558.

Conard, N. J. (2004). *Settlement Dynamics of the Middle Paleolithic and Middle Stone Age*. Kerns Verlag, Tübingen.

Cook, N. D. (2002). *Bihemispheric Language: How the Two Hemispheres Collaborate in the Processing of Language.*, pages 169–194. In Crow (2002).

Cooper, A. and Stringer, C. (2013). Did the Denisovans Cross Wallace's Line? *Science*, 342:321–323.

Coqueugniot, H., Hublin, J. J., Veillon, F. Houët, F., and Jacob, T. (2004). Early brain growth in *Homo erectus* and implications for cognitive ability. *Nature*, 431:299–302.

Corballis, M. C. (2002). *Laterality and Human Speciation*, pages 137–152. In Crow (2002).

Cosmides, L. and Tooby, J. (1994). Beyond intuition and instinct blindness: toward an evolutionary rigorous cognitive science. *Cognition*, 50:41–77.

Crockford, S. J. (2002). *Animal Domestication and Heterochronic Speciation: The Role of Thyoid Hormone*, pages 122–153. In Minugh-Purvis and McNamara (2002).

Crow, T. J. (2002). *The Speciation of Modern Homo Sapiens*. Oxford University Press, UK.

Crow, T. J. (2005). Who forgot Paul Broca? The origin of language as test case for speciation theory. *J. Linguistics*, 41:133–156.

Crow, T. J. (2006). March 27, 1827 and what happened later – the impact of psychiatry on evolutionary theory. *Progress in Neuro-Psychopharmacology & Biological Psychiatry*, 30:785–796.

Darwin, C. (1998 / 1872). *The Origin of Species*. The Modern Library, New York, 6th edition.

Davidson, I. and Noble, W. (1993). *Tools and language in human evolution*, pages 363–388. In Gibson and Ingold (1993).

Dawkins, R. (1991). *The Blind Watchmaker*. Penguin, UK.

de Ruiter, J. P. and Levinson, S. C. (2008). A biological infrastructure for communication underlies the cultural evolution of languages. *Behavioral and Brain Sciences*, 31:518.

Deacon, T. (1997). *The Symbolic Species*. Penguin, England.

Dean, C. (2000). Progress in Understanding Hominoid Dental Development. *Journal of Anatomy*, 197:77–101.

Dean, C., Leakey, M. G., Reid, D., Schrenk, F., Schwartz, G. T., Stringer, C., and Walker, A. (2001). Growth Processes in Teeth Distinguish Modern Humans from Homo Erectus and Earlier Hominins. *Nature*, 414:628–631.

Dean, M. C. (2006). Tooth microstructure tracks the pace of human life-history evolution. *Proc. R. Soc. B*, 273:2799–2808.

Deaner, R. O. and van Schaik, C. P. (2001). Flaws in evolutionary theory and interpretation: Commentary on Finlay et al. (2001). *Behavioral and Brain Sciences*, 24:282–283.

DeGraff, M. (2001). On the Origin of Creoles. *Linguistic Typology*, 5:213–310.

DeGusta, D. (1999). Hypoglossal Canal Size and Hominid Speech. *Proc. Natl. Acad. Sci. USA*, 96:1800.

Dennett, D. C. (1991). *Consciousness Explained*. Penguin, England.

Dennett, D. C. (1995). *Darwin's Dangerous Idea*. Penguin, London.

Dennett, D. C. (1998). *Brainchildren*. MIT Press, Cambridge, US.

d'Errico, F., Henshilwood, C., Lawson, G., Vanhaeren, M., Tillier, A., Soressi, M., Bresson, F., Maureille, B., Nowell, A., Lakarra, J., Blackwell, C., and Julien, M. (2003). Archaeological Evidence for the Emergence of Language, Symbolism, and Music – An Alternative Multidiciplinary Perspective. *Journal of World Prehistory*, 17:1–70.

Drayna, D. (2005). Founder Mutations. *Scientific American*, 293:60–67.

Dreyfus, H. L. (1979). *What Computers Still Can't Do*. MIT Press, Cambridge, Massachusetts.

Dunbar, R. I. M. (2001). Confounding explanation: Commentary on Finlay et al. (2001). *Behavioral and Brain Sciences*, 24:283.

Duret, L. (2009). Mutation patterns in the human genome: more variable than expected. *PLoS Biology*, 7 (2) e28.

Eckhardt, R. B. (2000). *Human Paleobiology*. Cambridge University Press.

Eldredge, N. (1996). *Reinventing Darwin: The Great Evolutionary Debate*. Phoenix, London.

Eldredge, N. (2003). *Human Triangles: Genes, Sex and Economics in Human Evolution*, pages 91–110. In Scher and Rauscher (2003).

Elliott, T. (2001). D'Arcy Wentworth Thompson, interindividual variation, and postnatal neuronal growth. *Behavioral and Brain Sciences*, 24:284.

Emmorey, K. (2005). Sign languages are problematic for a gestural origins theory of language evolution: commentary on Arbib 2005. *Behavioral and Brain Sciences*, 28:130–131.

Fabrega, H. J. (2005). Biological evolution of cognition and culture: Off Arbib's mirror-neuron system stage?: commentary on Arbib (2005). *Behavioral and Brain Sciences*, 28:131–132.

Falk, D. (1992). *Braindance*. Henry Holt and Co., New York.

Fernández-Jalvo, Y., Diéz, J. C., Cáceres, I., and Rosell, J. (1999). Human cannibalism in the Early Pleistocene of Europe (Gran Dolina, Sierrade Atapuerca, Burgos, Spain). *Journal of Human Evolution*, 37:591–622.

Finlay, B. L., Darlington, R. B., and Nicastro, N. (2001). Developmental structure in brain evolution. *Behavioral and Brain Sciences*, 24:263–308.

Fisher, S. E. (2006). Tangled webs: Tracing the connections between genes and cognition. *Cognition*, 101:270–297.

Fitch, W. T. (2005). Protomusic and protolanguage as alternatives to protosign: commentary on Arbib (2005). *Behavioral and Brain Sciences*, 28:132–133.

Fitch, W. T. and Hauser, M. D. (2004). Computational Constraints on Syntactic Processing in a Nonhuman Primate. *Science*, 303:377–380.

Fodor, J. A. (1983). *The Modularity of Mind*. MIT Press, England.

Fodor, J. A. (2008). Against Darwinism. *Mind & Language*, 23:1–24.

Fondon, J. W. and Garner, H. R. (2004). Molecular Origins of Rapid and Continuous Morphological Evolution. *Proc. Natl. Acad. Sci. USA*, 101:18058–18063.

Frayer, D. W. and Wolpoff, M. H. (1993). Reply to Milo and Quiatt (1993). *Current Anthropology*, 34:582–584.

Friederici, A. D. (2006). The Neural Basis of Language Development and Its Impairment. *Neuron*, 52:941–952.

Friederici, A. D., Bahlmann, J., Heim, S. Schubotz, R. I., and Anwander, A. (2006). The brain differentiates human and non-human grammars: Functional localization and structural connectivity. *PNAS*, 103 no. 7:2458–2463.

Friedman, W. (2009). The meaning of Darwin's "abominable mystery". *American Journal of Botany*, 96:5–21.

Gargett, R. H. (1989). Grave Shortcomings: The Evidence for Neandertal Burial. *Current Anthropology*, 30:157–190.

Gargett, R. H. (1993). Reply to Milo and Quiatt (1993). *Current Anthropology*, 34:584–585.

Gargett, R. H. (1999). Middle Palaeolithic burial is not a dead issue: the view from Qafzeh, Saint-Céaire, Kebara, Amud, and Dederiyeh. *Journal of Human Evolution*, 37:27–90.

Gargett, R. H. (2000). A response to Hovers, Kimbel and Rak's argument for the purposeful burial of Amud 7. *Journal of Human Evolution*, 39:261–266.

Gentner, T. Q., Fenn, K. M., Margoliash, D., and Nusbaum, H. C. (2006). Recursive syntactic pattern learning by songbirds. *Nature*, 440:1204–1207.

Gergely, G. and Csibra, G. (2005). A few reasons why we don't share Tomasello et al.'s intuitions about sharing: commentary on Tomasello et al. (2005). *Behavioral and Brain Sciences*, 28:701–702.

Gibson, K. R. and Ingold, T. (1993). *Tools, Language and Cognition in Human Evolution*. Cambridge University Press, Cambridge, UK.

Gilbert, S. F., Opitz, J. M., and Raff, R. A. (1996). Resynthesizing Evolutionary and Developmental Biology. *Developmental Biology*, 173:357–372.

Gilby, I. C., Thompson, M. E., Ruane, J. D., and Wrangham, R. (2010). No evidence of short-term exchange of meat for sex among chimpanzees. *Journal of Human Evolution*, 59:44–53.

Gilissen, E. and Simmons, R. M. T. (2001). Brain evolution: A matter of constraints and permissions? Commentary on Finlay et al. (2001). *Behavioral and Brain Sciences*, 24:284–286.

Gold, P. E. (1995). Role of glucose in regulating the brain and cognition. *American Journal of Clinical Nutrition*, 61:987S–995S.

Gould, S. J. (1977). *Ontogeny and Phylogeny*. Harvard University Press, Mass.

Gould, S. J. (1989). *Wonderful Life*. Penguin, UK.

Gould, S. J. (1994). Tempo and Mode in the Macroevolutionary reconstruction of Darwinism. *PNAS*, 91:6764–6771.

Gould, S. J. (2000). Of coiled oysters and big brains: how to rescue the terminology of heterochrony, now gone astray. *Evolution & Development*, 2:241–248.

Gould, S. J. and Eldredge, N. (1977). Punctuated equilibria: the tempo and mode of evolution reconsidered. *Paleobiology*, 11:2–12.

Gould, S. J. and Vrba, E. S. (1982). Exaptation: A Missing Term in the Science of Form. *Paleobiology*, 8:4–15.

Green, R., Malaspinas, A.-S., Krause, J., and Briggs, A. (2008). A Complete Neandertal Mitochondrial Genome Sequence Determined by High-Thoughput Sequencing. *Cell*, 134:416–426.

Green, R. E., Krause, J., Briggs, A. W., Maricic, T., Stenzel, U., Kircher, M., and Patterson, N. (2010). A draft sequence of the Neandertal genome. *Science*, 328:710–722.

Groves, C. P. (1989). *A Theory of Human and Primate Evolution*. Clarendon Press, Oxford, UK.

Haeckel, E. (1866). *Generelle Morphologie: Allgemeine Entwicklungsgeschichte der Organismen*. Georg Reimer, Berlin.

Haeusler, M. and McHenry, H. M. (2007). Evolutionary reversals of limb proportions in early hominds?: evidence from KNM-ER 3735 (Homo habilis). *Journal of Human Evolution*, 53:383–405.

Haile-Selassie, Y. (2001). Late Miocene Hominids from the Middle Awash, Ethiopia. *Nature*, 412:178–181.

Hall, B. K. (2002). *Evolutionary Development Biology: Where Embryos and Fossils Meet*, pages 7–27. In Minugh-Purvis and McNamara (2002).

Hallgrimsson, B., Lieberman, D. E., Liu, W., Ford-Hutchinson, A. F., and Jirik, F. R. (2007). Epigenetic interactions and the structure of phenotypic variation in the cranium. *Evolution and Development*, 9:76–91.

Hamai, M., Nishida, T., Takasaki, H., and Tjrner, L. A. (1992). New records of within-group infanticide and cannibalism in wild chimpanzees. *Primates*, 33:151–162.

Hauser, M. D., Barner, D., and O'Connell, T. (2007). Evolutionary linguistics: a new look at an old landscape. *Language, Learning and Development*, 3(2):101–132.

Hauser, M. D., Chomsky, N., and Fitch, W. T. (2002). The Faculty of Language: What is It, Who Has It, and How Did It Evolve? *Science*, 298:1569–1579.

Hayakawa, T., Angata, T., Lewis, A. L., Mikkelsen, T. S., Varki, N. M., and Varki, A. (2005). A Human-Specific Gene in Microglia. *Science*, 309:1693.

Henshilwood, C. S., d'Errico, F., van Niekerk, K. L., Coquinot, Y., Jacobs, Z., Lauritzen, S.-E., Menu, M., and García-Moreno, R. (2011). A 100,000-Year-Old Ochre-Processing Workshop at Blombos Cave, South Africa. *Science*, 334:219–222.

Heyes, C. and Huber, L. (2000). *The Evolution of Cognition*. MIT Press, Cambridge, Mass.

Hinzen, W. and Uriagereka, J. (2006). On The Metaphysics of Linguistics. *Erkenntnis*, 65:71–96.

Hlusko, L. J. (2004). Integrating the Genotype and Phenotype in Hominid Paleontology. *Proc. Natl. Acad. Sci. USA*, 101:2653–2657.

Horwitz, B., Husain, F. T., and Guenther, F. H. (2005). Auditory object processing and primate biological evolution: commentary on Arbib (2005). *Behavioral and Brain Sciences*, 28:134.

Hublin, J. J. and Coqueugniot, H. (2005). Absolute or proportional brain size: That is the question. *Journal of Human Evolution*, xx:1–5.

Hurford, J. R., Studdert-Kennedy, M., and Knight, C. (1998). *Approaches to the Evolution of Language*. Cambridge University Press, UK.

Indurkhya, B. (2005). On the neural grounding for metaphor and projection: commentary on Arbib (2005). *Behavioral and Brain Sciences*, 28:124–135.

Jablonka, E. and Raz, G. (2009). Transgenerational Epigenetic Inheritance: Prevalence, Mechanisms, and Implications for the Study of Heredity and Evolution. *The Quarterly Review of Biology*, 84:131–176.

Jablonski, N. G., Chaplin, G., and McNamara, K. J. (2002). *Natural Selection and the Evolution of Hominid Patterns of Growth and Development*, pages 189–206. In Minugh-Purvis and McNamara (2002).

Jackendoff, R. (1999). Possible Stages in the Evolution of the Language Capacity. *Trends in Cognitive Sciences*, 3:272–279.

Jackendoff, R. (2002). *Foundations of Language*. Oxford University Press, NY.

Jacobs, Z., Duller, G. A. T., Wintle, A. G., and Henshilwood, C. S. (2006). Extending the chronology of deposits at Blombos cave, South Africa, back to 140ka using optical dating of single and multiple grains of quartz. *Journal of Human Evolution*, 51:255–273.

James, S. R. (1989). Hominid Use of Fire in the Lower and Middle Pleistocene: A Review of the Evidence. *Current Anthropology*, 30:1–26.

Jarvis, E. D. (2004). *Learned Birdsong and the Neurobiology of Human Language*, pages 749–777. In Zeigler and Marler (2004).

Jenkins, L. (2000). *Biolinguistics*. Cambridge University Press, UK.

Jerrison, H. J. (1983). The Evolution of the Advanced Hominid Brain: reply to Blumenberg. *Current Anthropology*, 24:604–605.

Johanson, D. C. and White, T. D. (1979). A systematic assessment of early African hominids. *Science*, 203:321–330.

Johnson, L. L. (1989). The Neanderthals and Population as Prime Movers. *Current Anthropology*, 30:534–535.

Jolly, A. (1999). *Lucy's Legacy*. Harvard University Press, USA.

Jones, P. R. (1979). Effects of Raw Materials on Biface Manufacture. *Science*, 204:835–836.

Kalverboer, A., Hopkins, B., and Geuze, R. (1993). *Motor Development in Early and Later Childhood: Longitudinal Approaches*. Cambridge University Press.

Kaplan, J. T. and Iacoboni, M. (2005). Listen to my actions!: commentary on Arbib (2005). *Behavioral and Brain Sciences*, 28:135–136.

Kelso, J. (1995). *Dynamic Patterns*. MIT Press, Cambridge, Mass.

Kittler, R., Kayser, M., and Stoneking, M. (2003). Molecular Evolution of *Pendiculus humanus* and the Origin of Clothing. *Current Biology*, 13 Issue 16:1414–1417.

Klein, R. G. (2000). Archaeology and the Evolution of Human Behavior. *Evolutionary Anthropology*, 9:17–36.

Klinowska, M. (1994). *Brains, Behaviour and Intelligence in Cetaceans (Whales, Dolphins and Porpoises)*. The High North Publications.

Knight, C. (2000). *Play as Precursor of Phonology and Syntax*, pages 99–119. In Knight et al. (2000).

Knight, C., Studdert-Kennedy, M., and Hurford, J. (2000). *The Evolutionary Emergence of Language*. Cambridge University Press, Cambridge, UK.

Köhler, M. and Moyà-Solà, S. (1997). Ape-like or hominid-like? The ositional behavior of *Oreopithecus bambolii* reconsidered. *Proc. Natl. Acad.Sci. USA*, 94:11747–11750.

Kotchoubey, B. (2005). Pragmatics, prosody, and evolution: Language is more than a symbolic system: commentary on Arbib (2005). *Behavioral and Brain Sciences*, 28:136–137.

Krebs, P. R. (2007). Virtual Models and Simulations: A Different Kind of Science? *Techné: Research in Philosopy and Technology*, 11:42–54.

Krebs, P. R. (2008). *Smoke without Fire: What do Virtual Experiments in Cognitive Science Really Tell Us?*, chapter 13, pages 177–187. In Srinivasan et al. (2008).

Kunzig, R. (1997). Atapuerca: The Face of an Ancestral Child. *Discover*, 18:88–100.

Lahr, M. M. and Foley, R. (2004). Palaeoanthropology: Human Evolution Writ Small. *Nature*, 431:1043–1044.

Lanyon, S. J. (2005). *A "Sudden Appearance" Model for the Evolution for Human Cognition and Language*, pages 1248–1253. In Bara et al. (2005).

Lanyon, S. J. (2006). *A Saltationist Approach for the Evolution of Human Cognition and Language*, pages 176–183. In Cangelosi et al. (2006).

Leakey, R. (1992). *Origins Reconsidered.* Abacus, London.

Lee, M., Cau, A., Naish, D., and Dyke, G. (2014). Sustained miniaturization and anatomical innovation in the dinosaurian ancestors of birds. *Science*, 345:562–566.

Leonard, W. R. and Robertson, M. L. (1997). Rethinking the Energetics of Bipedality. *Current Anthropology*, 38:304–309.

Lewin, R. (1987). *Bones of Contention.* Simon and Schuster, NY.

Lieberman, D. E., McBratney, B. M., and Krovitz, G. (2002). The Evolution and Development of Cranial Form in Homo sapiens. *PNAS*, 99:1134–1139.

Lieberman, P. (1992). On Neanderthal Speech and Neanderthal Extinction. *Current Anthropology*, 33:409–410.

Lieberman, P., Reynolds, V., and Terrace, H. S. (1991). Apes and Us: An Exchange. *The New York Review of Books*, 38 No.16.

Lightfoot, D. (2000). *The Spandrels of the Linguistic Genotype*, pages 231–247. In Knight et al. (2000).

Lock, A. (1993). *Human language development and object manipulation*, pages 279–310. In Gibson and Ingold (1993).

Lordkipanidze, D., Ponce de León, M. S., Margvelashvili, A., Rak, Y., Rightmire, G. P., Vekua, A., and Zollikofer, C. P. E. (2013). A Complete Skull from Dmanisi, Georgia, and the Evolutionary Biology of Early Homo. *Science*, 342:326–331.

Lovejoy, C. O., Cohn, M. J., and White, T. D. (1999). Morphological Analysis of the Mammalian Postcranium: A Developmental Perspective. *Proc. Natl. Acad. Sci. USA*, 96:13247–13252.

MacLarnon, A. and Hewitt, G. (2004). Increased Breathing Control: Another Factor in the Evolution of Human Language. *Evolutionary Anthropology*, 13:181–197.

Macneilage, P. and Davis, B. (2000). *Evolution of Speech: The Relation Between Ontogeny and Phylogeny.*

Manzi, G., Arsuaga, A. G., and J., A. (2000). Cranial Discrete Traits in the Middle Pleistocene Humans from Sima de Los Huesos (Sierra de Atapuerca, Spain). *Journal of Human Evolution*, 38:425–446.

Marchal, F. (2000). A New Morphometric Analysis of the Hominid Pelvic Bone. *Journal of Human Evolution*, 38:347–365.

Marcus, G. F. (2006). Cognitive architecture and descent with modification. *Cognition*, 101:443–465.

Maresca, B. and Schwartz, J. H. (2006). Sudden Origins: A General Mechanism of Evolution Based on Stress Protein Concentration and Rapid Environmental Change. *The Anatomical Record*, 2898:38–46.

Margoliash, D. and Nusbaum, H. C. (2009). Language: the perspective from organismal bilogy. *trends in Cognitive Sciences*, 13:505–510.

Markson, L. and Diesendruck, G. (2005). Causal curiosity and the conventionality of culture: commentary on Tomasello et al. (2005). *Behavioral and Brain Sciences*, 28:709.

McBrearty, S. and Brooks, A. S. (2000). The revolution that wasn't: a new interpretation of the origin of modern human behavior. *Journal of Human Evolution*, 39:453–563.

McCollum, M. A. (1999). The Robust Australopithecine Face: A Morphogenetic Perspective. *Science*, 284:301–305.

McHenry, H. M. (1994). Tempo and mode in human evolution. *Proc. Natl. Acad. Sci.*, 91:6780–6786.

McKinney, M. (1991). *Heterochrony*. Plenum Press, NY.

McKinney, M. L. (1998). The Juvenilized Ape Myth - Our "Overdeveloped" Brain. *BioScience*, 48:109–116.

McKinney, M. L. (2002). *Brain Evolution by Stretching the Global Mitotic Clock of Development*, pages 173–188. In Minugh-Purvis and McNamara (2002).

McNeill, D., Bertenthal, B., Cole, J., and Gallagher, S. (2005). Gesture-first, but no gestures? *Behavioral and Brain Sciences*, 28:138–139.

Mellars, P. (2005). The Impossible Coincidence. A Single-Species Model for the Origins of Modern Human Behavior in Europe. *Evolutionary Anthropology*, 14:12–27.

Mellars, P. (2006). A new radiocarbon revolution and the dispersal of modern humans in Eurasia. *Nature*, 439:931–935.

Mendíl-Giró, J. L. (2006). *Language and Species. Limits and Scope of a Venerable Comparison*, pages 82–118. In Rosselló and Martín (2006).

Mercader, J., Panger, M., and Boesch, C. (2002). Excavation of a Chimpanzee Stone Tool Site in the African Rainforest. *Science*, 296:1452–1455.

Milo, R. G. and Quiatt, D. (1993). Glottogenesis and Anatomically Modern Homo Sapiens: The Evidence for and Implications of a Late Origin of Vocal Language. *Current Anthropology*, 34:569–598.

Milton, K. (2006). Diet and Primate Evolution. *Scientific American*, 16:22–29.

Minugh-Purvis, N. and McNamara, K. J. (2002). *Human Evolution Through Developmental Change*. The John Hopkins Press, USA.

Minugh-Purvis, N. and Radovčić (2000). Krapina 1: A Juvenile Neandertal From the Early Late Pleistocene of Croatia. *American Journal of Physical Anthropology*, 111:393–424.

Mitteroecker, P., Gunz, P., Bernhard, M., Schaefer, K., and Bookstein, F. L. (2004). Comparison of cranial ontogenetic trajectories among Great Apes and humans. *Journal of Human Evolution*, 46:679–698.

Montagu, A. (1989). *Growing Young*. Bergin and Garvey, Mass. USA.

Moore, R. and Moore, J. (2006). *Evolution 101*. Greenwood Press, US.

Moyà-Solà, S., Köhler, M., and Rook, L. (1999). Evidence of hominid-like precision grip capability in the hand of the Miocene ape *Oreopithecus*. *Proc. Natl. Acad. Sci. USA*, 96:313–317.

Mundale, J. (2003). *Evolutionary Psychology and the Information-Processing Model of Cognition*, pages 229–241. In Scher and Rauscher (2003).

Muotri, A. R. and Gage, F. H. (2006). Generation of neuronal variability and complexity. *Nature*, 44:1087–1093.

Naef, A. 1926, Z. M. u. S. d. A. N. .-. (1926). Zur Morphologie und Stammesgeschichte des Affenschaedels. *Naturwissenshaften*, 14:89–97.

Newman, S. A. and Bhat, R. (2008). Dynamical patterning modules: physico-genetic determinants of morphological development and evolution. *Physical Biology*, 5:15008–15014.

Newman, S. A. and Müller, G. B. (2005). Origination and Innovation in the Vertebrate Limb Skeleton: An Epigenetic Perspective. *Journal of Experimental Zoology (Mol Dev Evol)*, 304B:593–609.

Nishimura, T., Mikami, A., Suzuki, J., and Matsuzawa, T. (2006). Descent of the hyoid in chimpanzees: evolution of face flattening and speech. _Journal of Human Evolution_, 51:244–254.

Noble, J. (2000). _Cooperation, Competition and the Evolution of Prelinguistic Communication_, pages 40–61. In Knight et al. (2000).

Noble, W. and Davidson, I. (1997). _Human Evolution, Language and Mind_. Cambirdge University Press, UK.

Oliver, A., Johnson, M. H., Karmiloff-Smith, A., and Pennington, B. (2000). Deviations in the emergence of representations: a neuroconstructivist framework for analysing developmental disorders. _Developmental Science_, 3:1–40.

Owren, M. J. and Rendall, D. (2001). Sound on the Rebound: Bringing Form and Function Back to the Forefront in Understanding Nonhuman Primate Vocal Signaling. _Evolutionary Anthropology_, 10:58–71.

Pagel, M. (2000). _The History, Rate and Pattern of World Linguistic Evolution_, chapter 22, pages 391–416. In Knight et al. (2000).

Pagni, L. and Baccetti, T. (1993). Heredity and environment in the genesis, epigenesis and evolution of the orofacial area. _Minerva Stomatol_, 42(102):1–13.

Palmer, A. R. (2004). Symmetry breaking and the evolution of development. _Science_, 306:828–833.

Panger, M. A., Brooks, A. S., Richmond, B. G., and Wood, B. (2002). Older Than the Oldowan? Rethinking the Emergence of Hominin Tool Use. _Evolutionary Anthropology_, 11:235–245.

Panksepp, J. (1998). _Affective Neuroscience_. Oxford University Press, New York.

Parisi, D. (2003). *Evolutionary Psychology and Artificial Life*, pages 243–265. In Scher and Rauscher (2003).

Parker, S. T. and Milbrath, C. (1993). *Higher intelligence, propositional language, and culture as adaptations for planning*, pages 314–333. In Gibson and Ingold (1993).

Patterson, N., Richter, D. J., Gnerre, S., Lander, E. S., and Reich, D. (2006). Genetic evidence for complex speciation of humans and chimpanzees. *Nature*, 44:1103–1108.

Pearson, O. M. (2004). Has the Combination of Genetic and Fossil Evidence Solved the Riddle of Modern Human Origins? *Evolutionary Anthropology*, 13:145–159.

Penn, D. C., Holyoak, K. J., and Povinelli, D. J. (2008). Darwin's mistake: Explaining the discontinuity between human and nonhuman minds. *Behavioral and Brain Sciences*, 31:109–178.

Peters, E. H. (1993). Reply to Milo and Quiatt (1993). *Current Anthropology*, 34:587–588.

Petitto, L. A. and Marentette, P. F. (1991). Babbling in the Manual Mode: Evidence for the Ontogeny of Language. *Science*, 251:1493–1496.

Phillips, B. L. and Shine, R. (2006). An invasive species induces rapid adaptive change in a native predator: cane toads and black snakes in Australia. *Proc. R. Soc. B*, 273:1545–1550.

Pi, S. J., Vea, J. J., and Serrallonga, J. (1997). Did the First Hominids Build Nests? *Current Anthropology*, 38:914–916.

Piattelli-Palmarini, M. (1989). Evolution, Selection and Cognition: From "Learning" to Parameter Setting in Biology and the Study of Language. *Cognition*, 31:1–44.

Piattelli-Palmarini, M., Hancock, R., and Bever, T. (2008). Language as ergonomic perfection. *Behavioral and Brain Sciences*, 31:530–531.

Pickering, T. R., White, T. D., and Toth, N. (2000). Cutmarks on a Plio-Pleistocene hominid from Sterkfontein, South Africa. *American Journal of Physical Anthropology*, 111:579–584.

Pigliucci, M. and Kaplan, J. (2006). *Making Sense of Evolution.* The University of Chicago Press.

Pika, S., Liebel, K., and Tomasello, M. (2005). Gestural Communication in Subadult Bonobos (Pan Paniscus): Repertoire and Use. *American Journal of Primatology*, 65:39–61.

Pinker, S. (1994). *The Language Instinct.* William Morrow & Co, NY.

Pinker, S. (1997). *How The Mind Works.* Penguin, Australia.

Pinker, S. and Bloom, P. (1990). Natural Language and Natural Selection. *Behavioral and Brain Sciences*, 13:707–727.

Plavcan, J. M. (1997). Interpreting hominid behavior on the basis of sexual dimorphism. *Journal of Human Evolution*, 32:345–374.

Ploog, D. (2002). *Is the Neural Basis of Vocalisation Different in Non-Human Primates and Homo Sapiens?*, pages 121–135. In Crow (2002).

Pollack, R. (1994). *Signs of Life.* Houghton Mifflin, Boston, US.

Povinelli, D. J. and Barth, J. (2005). Reinterpreting behavior: A human specialization?: commentary on Tomasello et al. (2005). *Behavioral and Brain Sciences*, 28:712–713.

Povinelli, D. J. and Vonk, J. (2003). Chimpanzee minds: suspiciously human? *Trends in Cognitive Sciences*, 7:157–160.

Provine, R. R. (2005). Illusions of intentionality, shared and unshared: commentary on Tomasello et al. (2005). *Behavioral and Brain Sciences*, 28:713–714.

Pruetz, J. D. and Bertolani, P. (2007). Savanna Chimpanzees, *Pan troglodytes verus*, Hunt with Tools. *Current Biology*, 17:412–417.

Quartz, S. R. (2003). *Toward a Developmental Evolutionary Psychology: Genes, Development, and the Evolution of the Human Cognitive Architecture*. In Scher and Rauscher (2003).

Rae, C., Hare, N., Bubb, W. A., McEwan, S. R., Bröer, A., McQuillan, J. A., Balcar, V. J., Conigrave, A. D., and Bröer, S. (2003). Inhibition of glutimine transport depletes glutamate and GABA neurotransmitter pools: futher evidence for metabolic compartmentation. *Journal of Neurochemistry*, 85:503–514.

Raff, R. A. (1996). *The shape of life: genes, development, and the evolution of animal form*. University of Chicago Press, Chicago.

Raff, R. A. (2000). Evo-devo: the evolution of a new discipline. *Nature reviews genetics*, 1:74–79.

Rakic, P. (1995). A Small Step for the Cell, a Giant Leap for Mankind: A Hypothesis of Neocortical Expansion During Evolution. *Trends in Neuroscience*, 18:383–388.

Ramus, F. (2006). Genes, brain, and cognition: A roadmap for the cognitive scientist. *Cognition*, 101:247–269.

Ranov, V. A., Carbonell, E., and Rodriguez, X. P. (1995). Kuldara: Earliest Human Occupation in Central Asia in its Afro-Asian Context. *Current Anthropology*, 36:337–346.

Rauschecker, J. (2005). Vocal gestures and auditory objects: commentary on Arbib (2005). *Behavioral and Brain Sciences*, 28:143–144.

Reid, R. G. B. (2007). *Biological Emergences: Evolution by Natural Experiment*. The MIT Press USA.

Reisz, R. R., Scott, D., Sues, H., Evans, D. C., and Raath, A. (2005). Embryos of an Early Jurassic Prosauropod dinosaur and Their Evolutionary Significance. *Science*, 309:761–764.

Renfrew, C. and Bahn, P. (1994). *Archaeology Theories Methods and Practice*. Thames and Hudson, London.

Richardson, M. K. and Keuck, G. (2002). Haeckel's ABC of evolution and development. *Biological Review*, 77:495–528.

Richerson, P. J. and Boyd, R. (2000). *Climate, Culture, and the Evolution of Cognition*, pages 329–346. In Heyes and Huber (2000).

Richmond, B. G. (1999). Eurasian Hominoid Evolution. *Evolutionary Anthropology*, 7:194–196.

Rightmire, G. P. (1990). *The Evolution of Homo Erectus*. Cambridge University Press, Uk.

Rightmire, P. G. (1996). The human cranium from Bodo, Ethiopia: evidence for speciation in the Middle Pleistocene? *Journal of Human Evolution*, 31:21–39.

Rilling, J. K. and Seligman, R. A. (2002). A Quantitative Morphometric Comparative Analysis of the Primate Temporal Lobe. *Journal of Human Evolution*, 42:505–533.

Rizzolatti, G., Fadiga, L., Fogassi, L., and Gallese, V. (1996). Premotor cortex and the recognition of motor actions. *Cognitive Brain Research*, 3:131–141.

Robert, J. S. (2006). *Embryology, Epigenesis, and Evolution: Taking Development Seriously*. Cambridge University Press.

Robertson, W. R. B. (1916). Chromosome studies. 1. Taxonomic relationships shown in the chromosomes of *Tettigidae* and *Acrididae*. V-shaped chromosomes and their significance in *Acrididae*, *Locustidae* and *Grylidae*: chromosome and variation. *Journal of Morphology*, 27:179–331.

Rosselló, J. and Martín, J. (2006). *The Biolinguistic Turn.* PPU, Promociones y Publicaciones Universitarias, Barcelona.

Rozzi, F. V., d'Errico, F., Vanhaeren, M., Grootes, P. M., Kerauret, B., and Dujardin, V. (2009). Cutmarked human remains bearing Neanderthal features and modern human remains associated with the Aurignacian at Les Rois. *J Anthropological Science*, 87:153–185.

Ruff, C. B., Trinkaus, E., and Holliday, T. W. (1997). Body mass and encephalization in Pleistocene *Homo. Nature*, 387:173–176.

Savage-Rumbaugh, E. S. and Wilkerson, B. J. (1978). Socio-sexual behavior in *Pan paniscus* and *Pan troglodytes*: A Comparative Study. *Journal of Human Evolution*, 7:327–344.

Savage-Rumbaugh, S. (1994). *Kanzi.* John Wiley & Sons, US.

Schenker, N. M., Desgouttes, A., and Semendeferi, K. (2005). Neural connectivity and cortical substrates of cognition in hominoids. *Journal of Human Evolution*, 49:547–569.

Scher, S. and Rauscher, M. (2003). *Evolutionary Psychology: Alternative approaches.* Kluwer.

Schoenemann, P. T. (2006). Evolution of the Size and Functional Areas of the Human Brain. *Annual Review of Anthropology*, 35:379–406.

Schoenemann, P. T., Sheehan, M. J., and Glotzer, L. D. (2005). Prefrontal White Matter Volume is Disproportionately Larger in Humans Than in Other Primates. *Nature Neuroscience*, 8:242–252.

Schuster, R. (2005). Why not chimpanzees, lions, and hyenas too?: commentary on Tomasello et al. (2005). *Behavioral and Brain Sciences*, 28:716717.

Schwartz, J. H. (1999). *Sudden Origins: Fossils, Genes, and the Emergence of Species.* John Wiley & Sons, USA.

Schwartz, J. H. (2001). Adaption and Evolution. _Hist. Phil. Life Sci._, 23:505–517.

Schwartz, J. H. and Maresca, B. (2006). Do Molecular Clocks Run at All? A Critique of Molecular Systematics. _Biological Theory_, 1(4):357–371.

Semaw, S. (2000). The World's Oldest Stone Artifacts from Gona, Ethiopia: Their Implications for Understanding Stone Technology and Patterns of Human Evolution Between 2.6-1.5 Million Years Ago. _Journal of Archaeological Science_, 27:1197–1214.

Serre, D., Langaney, A., Chech, M., Teschler-Nicola, M., Paunovic, M., and Mennecier, P. (2004). No evidence of Neandertal mtDNA contribution to early modern humans. _PLoS Biology_, 2:313–317.

Smith, B. H. and Tompkins, R. L. (1995). Toward a Life History of the Hominidae. _Annual review of Anthropology_, 24:257–279.

Snowdon, C. T. (1993). _A comparative approach to language parallels_, pages 109–128. In Gibson and Ingold (1993).

Soressi, M. (2004). _From The Mousterian of Acheulian Tradition Type A to Type B: A Change in Technical Tradition, Raw Material, Task, or Settlement Dynamics?_, pages 343–366. In Conard (2004).

Sperber, D. and Wilson, D. (1995). _Relevance._ Blackwell, UK.

Srinivasan, N., Gupta, A. K., and Pandey, J. (2008). _Advances in Cognitive Science._ Sage, New Delhi.

Stahl, A. B. (1989). Commentary on James (1989): Hominid Use of Fire in the Lower and Middle Pleistocene: A Review of the Evidence. _Current Anthropology_, 30:18–19.

Steele, J. (2002). _When Did Directional Asymmetry Enter the Record?_, pages 153–168. In Crow (2002).

Sterelny, K. (2003). *Thought in a Hostile World*. Blackwell, UK.

Stern, J. T. (2000). Climbing to the Top: A Personal Memoir of Australopithecus Afarensis. *Evolutionary Anthropology*, 9:113–133.

Stiner, M. C., Munro, N. O., Surovell, T. A., Tchernov, E., and Bar-Yosef, O. (1999). Paleolithic Population Growth Pulses Evidenced by Small Animal Exploitation. *Science*, 283:190–197.

Stotz, K. C. and Griffiths, P. E. (2003). *Dancing in the Dark: Evolutionary Psychology and the argument from Design*, pages 135–160. In Scher and Rauscher (2003).

Stout, D., Toth, N., and Schick, K. (2000). Stone Tool-Making and Brain Activation: Positron Emission Tomography (PET) Studies. *Journal of Archaeological Science*, 27:1215–1223.

Stringer, C. (2003). Out of Ethiopia. *Nature*, 423:692–694.

Stringer, C. and Gamble, C. (1993). *In Search of the Neanderthals: Solving the Puzzle of Human Origins*. Thames and Hudson, London.

Studdert-Kennedy, M. (2000). *Emergence of Phonetic Structure*, pages 123–129. In Knight et al. (2000).

Suddendorf, T. and Busby, J. (2003). Mental time travel in animals. *Trends in Cognitive Sciences*, 7:391–396.

Tardieu, C. (1998). Short Adolescence in Early Hominids: Infantile and Adolescent Growth of the Human Femur. *American Journal of Physical Anthropology*, 107:163–178.

Tattersall, I. (2001). How We Came to Be Human. *Scientific American*, December:56–63.

Tattersall, I. (2006). How We Came to Be Human. *Scientific American*, 16:66–73.

Tattersall, I. and Schwartz, J. H. (1998). Morphology, Paleoanthropology, and Neanderthals. *The Anatomical Record*, pages 113–117.

Thompson, R. (2000). Categorical Perception and Conceptual Judgments by Nonhuman Primates. *Cognitive Science*, 24:363–396.

Tobias, P. V. (1998). Ape-Like Australopithecus after Seventy Years: Was it a Hominid? *Journal of the Royal Anthropological Institute*, 4:283–308.

Tocheri, M. W., Orr, C. M., Larson, S. G., Sutikna, J., Saptomo, E. W., Due, R. A., Djubiantono, T., Morwood, M. J., and Jungers, W. L. (2007). The Primitive Wrist of *Homo floresiensis* and Its Implications for Hominin Evolution. *Science*, 317:1743–1745.

Tomasello, M. (2000a). Do young children have adult syntactic competence? *Cognition*, 74:209–253.

Tomasello, M. (2000b). *Two Hypotheses About Primate Cognition*, pages 165–183. In Heyes and Huber (2000).

Tomasello, M., Call, J., and Gluckman, A. (1997). Comprehension of Novel Communicative Signs by Apes and Human Children. *Child Development*, 68:1067–1080.

Tomasello, M., Call, J., and Hare, B. (2003). Chimpanzees understand psychological states - the question is which ones and to what extent. *Trends in Cognitive Sciences*, 7:153–156.

Tomasello, M., Carpenter, M., Call, J., Behne, T., and Moll, H. (2005). Understanding and sharing intentions: The origins of cultural cognition. *Behavioral and Brain Sciences*, 28:675–735.

Toth, N. and Schick, K. (1993). *Early stone industries and inferences regarding language and cognition*, pages 346–362. In Gibson and Ingold (1993).

Tremblay, M., Lowery, R. L., and Majewska, A. K. (2010). Microglial Interactions with Synapses Are Modulated by Visual Experience. *PLoS Biology*, 8 (11) doi:10.1371/journal.pbio.10000527.

Trut, L., Oskina, I., and Kharlamova, A. (2009). Animal evolution during domestication: the domesticated fox as a model. *BioEssays*, 31:349–360.

Ulbaek, I. (1998). *The Origin of Language and Cognition*, pages 30–43. In Hurford et al. (1998).

Valladas, H., Reyss, J. L., Joron, J. L., Valladas, G., Bar-Yosef, O., and Vandermeersch, B. (1988). Thermoluminescence dating of Mousterian 'Proto-Cro-Magnon' remains from Israel and the origin of modern man. *Nature*, 331:614–616.

Van Schaik, C. (2006). Why are some animals so smart? *Scientific American*, 16:30–37.

Vanhaeren, M., d'Errico, F., Stringer, C., James, S. L., Todd, J. A., and Mienis, H. K. (2006). Middle Paleolithic Shell Beads in Israel and Algeria. *Science*, 312:1785–1788.

Vihman, M. M. and DePaolis, R. A. (2000). *The Role of Mimesis in Infant Language Development: Evidence for Phylogeny?*, pages 130–145. In Knight et al. (2000).

Vinicius, L. (2005). Human encephalization and developmental timing. *Journal of Human Evolution*, 49:762–776.

Walker, A. and Shipman, P. (1996). *The Wisdom of the Bones*. Alfred A. Knopf, NY.

Wallace, R. (1989). Cognitive Mapping and the Origins of Language and Mind. *Current Anthropology*, 30:518–526.

Wallace, R. (2005). A Global Workspace perspective on mental disorders. *Theoretical Biology and Medical Modelling*, 2:49.

Watson, J. S. (2005). "Einstein's baby" could infer intentionality: commentary on Tomasello et al. (2005). *Behavioral and Brain Sciences*, 28:719–720.

Weiner, S., Qinqi., X., Goldberg, P., Liu., J., and Bar-Josef, O. (1998). Evidence for the Use of Fire at Zhoukoudian, China. *Science*, 281:251–253.

West-Eberhard, M. J. (2005). Phenotypic accommodation: adaptive innovation due to developmental plasticity. *Journal of experimental zoology*, 304B:610–618.

Wilson, M. L., Wallauer, W. R., and Pusey, A. E. (2004). New Cases of Intergroup Violence Among Chimpanzees in Gombe National Park, Tanzania. *International Journal of Primatology*, 25:523–549.

Wong, K. (2003). An Ancestor To Call Our Own. *Scientific American*, Vol 288:42–51.

Wray, A. (2005). The explanatory advantages of the holistic protolanguage model: The case of linguistic irregularity: commentary on Arbib (2005). *Behavioral and Brain Sciences*, 28:147–148.

Wuethrich, B. (1998). Geological Analysis Damps Ancient Chinese Fires. *Science*, 281:165–166.

Wynn, T. (1993). *Layers of thinking in tool behavior*, pages 389–406. In Gibson and Ingold (1993).

Zeigler, H. P. and Marler, P. (2004). *Behavioral Neurobiology of Birdsong*. The New York Academy of Sciences, New York.

Zhang, K. and Sejnowski, T. J. (2000). A universal scaling law between gray matter and white matter of cerebral cortex. *PNAS*, 97:5621–5626.

Zlatev, J., Persson, T., and Gärdenfors, P. (2005). Triadic bodily mimesis is the difference: commentary on Tomasello et al. (2005). *Behavioral and Brain Sciences*, 28:720–721.

www.ingramcontent.com/pod-product-compliance
Lightning Source LLC
Chambersburg PA
CBHW070313190526
45169CB00005B/1605